# Introduction to Organic Spectroscopy

## Laurence M. Harwood

Professor and Head of Organic Chemistry, University of Reading

## Timothy D.W. Claridge

NMR Facility Manager, The Dyson Perrins Laboratory, University of Oxford

**OXFORD**
UNIVERSITY PRESS

# OXFORD

UNIVERSITY PRESS

Great Clarendon Street, Oxford OX2 6DP

Oxford University Press is a department of the University of Oxford.
It furthers the University's objective of excellence in research, scholarship,
and education by publishing worldwide in

Oxford  New York

Athens  Auckland  Bangkok  Bogotá  Buenos Aires  Calcutta
Cape Town  Chennai  Dar es Salaam  Delhi  Florence  Hong Kong  Istanbul
Karachi  Kuala Lumpur  Madrid  Melbourne  Mexico City  Mumbai
Nairobi  Paris  São Paulo  Singapore  Taipei  Tokyo  Toronto  Warsaw

with associated companies in  Berlin  Ibadan

Oxford is a registered trade mark of Oxford University Press
in the UK and in certain other countries

Published in the United States
by Oxford University Press Inc., New York

First published 1997
Reprinted 1999, 2000, 2001, 2002

A catalogue record for this book is available from the British Library

Library of Congress Cataloging in Publication Data
(Data available)
ISBN 0 19 855755 8

Printed in Great Britain
on acid-free paper
The Bath Press

# Contents

# Founding Editor's Foreword

The deduction of chemical structure from spectroscopic data is an essential skill required of all organic chemists, which is acquired mainly by the practice of interpreting original data. Laurence Harwood and Tim Claridge provide in this primer the basic knowledge, in the four most important areas of spectroscopy, which will allow students to start interpreting their own data and hence gain this essential skill.

Oxford Chemistry Primers have been designed to provide concise introductions relevant to all students of chemistry and contain only the essential material that would normally be covered in an 8–10 lecture course. In this primer the authors introduce students to this enormous area of basic structural analysis in a logical and easy to follow fashion. This primer will be of interest to apprentice and master chemist alike.

Stephen G. Davies
*The Dyson Perrins Laboratory*
*University of Oxford*

# Preface

The need to determine the structures of molecules efficiently lies at the heart of chemistry and the great strides made in the various techniques of spectroscopic analysis have been the root cause of the explosive burgeoning, particularly of organic chemistry, during the latter half of this century. In response to the chemist's desire to analyse ever smaller amounts of structures of increasing complexity, the advance of instrumentation and spectroscopic techniques shows no sign of slowing. To the newcomer, the task of selecting a starting point within the bewildering array that falls under the heading of 'spectroscopy' must appear very daunting indeed. The aim of this book is to provide a straightforward introduction to the basic principles, operation, and analyses possible using four of the spectroscopic techniques most routinely used by organic chemists—ultraviolet, infrared and nuclear magnetic resonance spectroscopy, and mass spectrometry.

Our thanks go to Dr Robin Aplin for helpful comments on the chapter on mass spectometry and Dr Katya Vines for proof-reading the manuscript.

We would like to dedicate this book to the memory of Andy Derome, our friend and colleague, whose tragic and untimely death at the age of 33 in 1991 cut short the career of someone who had established himself as one of the leading exponents of NMR spectroscopy in the world at the time. We are all the poorer for his passing away.

*Oxford, Reading*
October 1996

L. M. H. and T. D. W. C.

# 1 Introductory theory

## 1.1 The electromagnetic spectrum

The development of non-destructive spectroscopic methods of analysis which can be carried out on small amounts of material has provided the fundamental thrust behind the burgeoning of organic chemistry during the latter half of this century. Identification of unknown molecules of high complexity can now be carried out more or less routinely on samples ranging from several nanograms to a milligram of material, and in many cases it may not even be necessary to obtain pure samples.

Four techniques are used routinely by organic chemists for structural analysis. *Ultraviolet spectroscopy* was the first to come into general use during the 1930s. This was followed by *infrared spectroscopy* in the 1940s, with the establishment of *nuclear magnetic resonance spectroscopy* and *mass spectrometry* during the following two decades. Of these, the first three fall into the category of *absorption spectroscopy*. As this term suggests, these analytical techniques involve absorption of specific energies of *electromagnetic radiation* which correspond exactly in energy to specific excitations within the molecule being examined. Generally therefore, it follows that knowledge of the transitions which may be induced by absorption of a certain wavelength of electromagnetic radiation can be used to infer structural features.

In the classical treatment, electromagnetic radiation can be considered as a propagating wave of electrical energy with an orthogonal magnetic component oscillating with exactly the same frequency (Fig. 1.1).

Following this approach, electromagnetic radiation can be described by either its *frequency* or *wavelength*. These values are inversely proportional to each other being related by the following equation:

$$\lambda v = c$$

$\lambda$ = wavelength of the radiation
$v$ = frequency of the electromagnetic radiation
$c$ = speed of light $3 \times 10^8 \ ms^{-1}$

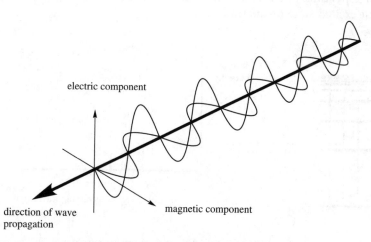

electric component

direction of wave propagation

magnetic component

**Fig. 1.1** The classical wave form depiction of electromagnetic radiation.

The shorter the wavelength (or higher the frequency) of electromagnetic radiation, the higher its energy and *vice versa*. Organic molecules absorb different wavelengths of electromagnetic radiation and undergo energetic transitions as a consequence of energy transfer. However, the classical treatment does not give us the full picture of the energy transfer process. In the quantum mechanical description, electromagnetic radiation can be considered to be propagated in discrete packets of energy called **photons**. These photons possess very specific energies and are said to be **quantized**. The energy of each photon is derived by the relationship:

$$E = h\nu$$

$h$ = Planck's constant $6.626 \times 10^{-34}$ Js

This description, involving quanta of energy makes it easier to appreciate the fact that transfer of energy does not occur over a continuous range, but only if the energy being transferred exactly matches that of the transition being carried out—in other words if the energy of the quantum of electromagnetic radiation corresponds exactly to the energy of the transition. Thus, excitation of an organic molecule involves absorption of specific quanta of electromagnetic radiation. This quantized energy is transferred to the molecule causing it to become promoted to a higher energy level, with the exact nature of the excitation depending upon the amount of electromagnetic energy absorbed. However, in any event, the photon of electromagnetic radiation will only be absorbed by a molecule if the energy it possesses corresponds *exactly* to an energy difference between two states of the molecule. This relationship is described by a simple extension of the above equation:

$\Delta E$ = the difference between the higher and lower states of the absorbing species.

$$\Delta E = h\nu$$

While, in theory, any photon corresponding to any possible transition within the molecule might be absorbed, the vast majority of the molecules being irradiated will exist initially in the unexcited **ground state** and the easiest, most efficient, absorption process will be that involving the lowest energy input, elevating the molecule to its **first excited state**. Hence the strongest absorption of electromagnetic radiation will occur at an energy corresponding to transitions from the molecule in its ground state to the first excited state of the transition.

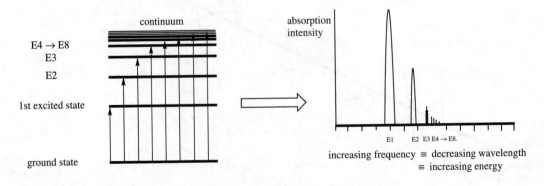

**Fig. 1.2**  An idealized representation of excitation transitions and how these are translated into an absorption spectrum.

It will be seen from this idealized and very simplified consideration that absorption spectra provide two kinds of information which can enable the electronic and structural characteristics of the molecule to be unravelled. Firstly, the ***absorption wavelength*** or ***frequency*** gives the energy associated with a particular excitation which can be related to the functional group responsible for absorption. Secondly, the ***absorption intensity*** reflects both the ease of the transition (which once again can provide information about the functionality undergoing excitation) and the concentration of the absorbing species.

Organic chemists often calculate the absorption intensity in ultraviolet spectroscopy (Chapter 2), but largely ignore it in infrared spectra (Chapter 3) other than a cursory description of absorptions as 'strong' or 'weak'. In proton nuclear magnetic resonance spectroscopy however, signal intensity is a key piece of information relating to the relative number of protons in the molecule responsible for the signal (Chapters 4 and 5).

## 1.2   Absorption spectroscopy

The transitions effected on absorbing a photon of a given wavelength are summarized in Fig. 1.3 and the regions of most interest to organic chemists for spectroscopic analysis are highlighted.

Increasing energies of electromagnetic radiation may cause ***rotational***, ***vibrational***, and ***electronic*** transitions to occur within a molecule and some of these frequencies of the electromagnetic spectrum are of greater utility to organic chemists than others for spectroscopic analysis.

$\gamma$-Rays and X-rays are so energetic that they simply cause electrons to be ejected from the molecule resulting in bond rupture, so their absorption provides little detailed structural information that could be of use to organic chemists (X-ray *diffraction*, of course, is a very powerful tool for structural analysis but is a quite separate technique). Similarly the vast number of possible rotational excitations which exist within organic molecules of the structural complexity usually encountered means that the far infrared and microwave regions of the electromagnetic spectrum do not usually provide the organic chemist with useful information.

Electromagnetic radiation in the ultraviolet and visible regions of the spectrum is sufficiently energetic to cause bonding electrons to be excited into higher energy orbitals. In organic molecules, this is a diagnostic feature for ***conjugated*** double bonds where the delocalization of $\pi$-electrons over

Ultraviolet / visible radiation causes electronic excitation.

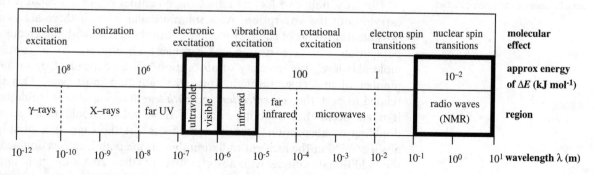

| nuclear excitation | ionization | | electronic excitation | vibrational excitation | | rotational excitation | | electron spin transitions | nuclear spin transitions | **molecular effect** |
|---|---|---|---|---|---|---|---|---|---|---|
| $10^8$ | | $10^6$ | | | | 100 | | 1 | $10^{-2}$ | **approx energy of $\Delta E$ (kJ mol$^{-1}$)** |
| $\gamma$–rays | X–rays | far UV | ultraviolet | visible | infrared | far infrared | microwaves | | radio waves (NMR) | **region** |
| $10^{-12}$   $10^{-10}$   $10^{-9}$   $10^{-8}$ | | | $10^{-7}$ | $10^{-6}$ | $10^{-5}$ | $10^{-4}$   $10^{-3}$ | $10^{-2}$   $10^{-1}$ | | $10^0$ | $10^1$ **wavelength $\lambda$ (m)** |

**Fig. 1.3**   The electromagnetic spectrum. **Note** the wavelength scale is logarithmic.

the whole system causes a decrease in the energy gap between the **highest occupied molecular orbital (HOMO)** and the **lowest unoccupied molecular orbital (LUMO)**. Increasing conjugation permits excitation by progressively lower frequency (longer wavelength) radiation (Section 2.1). Structural features of organic molecules which lead to absorption in the ultraviolet and visible region are often referred to as **chromophores**.

The lower frequency infrared radiation cannot promote electronic excitation, but is sufficiently energetic to cause bonds to deform. The energy required to stretch or bend a bond depends upon its force constant and this is in turn dependent upon the constituent atoms and whether they are singly or multiply bonded. Consequently, infrared spectroscopy provides insight into the construction of organic molecules without the prerequisite for conjugation.

Undeniably the electromagnetic radiation of most use to the organic chemist lies in the radiofrequency region of the spectrum. Here the extremely low energy radiation is just right to cause nuclei of elements having a **magnetic moment** to undergo transitions between nuclear spin states when they are placed in a strong magnetic field and the technique is termed nuclear magnetic resonance spectroscopy. Only nuclei which possess **nuclear spin** can be detected using this technique and the commonest nucleus with this property in organic molecules is the hydrogen nucleus. Proton nuclear magnetic resonance spectroscopy is the single most important analytical tool available to the organic chemist. Although the major isotope of carbon ($^{12}C$) does not possess nuclear spin the $^{13}C$ isotope, constituting 1.11% of natural carbon, does have spin and provides the means of investigating the carbon framework of molecules. However, the low natural abundance of the isotope, together with the low sensitivity of the technique for this nucleus means that $^{13}C$ NMR spectroscopy has only become routine with the greater sensitivity available with **Fourier transform** spectrometers (Section 1.3).

### Absorption intensity

Three factors affect the intensity of absorption of electromagnetic radiation. The **transition probability** in its most general definition is a measure of the likelihood that any specific transition will take place and is commonly simplified into **allowed** and **forbidden** transitions. Consideration of the theory is beyond the scope of this book, but we will come across allowed and forbidden transitions in ultraviolet spectroscopy.

The remaining two factors reflect the quantities of species capable of carrying out the absorption. At a sub-molecular level, if there are two possible transitions of equal probability, then the species with the greatest population will give rise to the strongest absorption. At a supra-molecular level, the intensity of absorption will be dependent upon the number of molecules through which the radiation must pass. This is related to both the **concentration** and **path length** of the sample. Initially, it might seem that doubling either of these values would result in a doubling of absorption intensity but this is not the case. If a sample absorbs 50% of the incident radiation then, if the path length is doubled, the additional sample will absorb 50% of the remaining radiation arriving at it. However, this corresponds to only 50% of the original

**Chromophores** result from the presence of conjugation.

Infrared radiation causes vibrational excitation.

Radio frequency radiation causes nuclear spin transitions.

The term **forbidden** in this context actually means strongly disfavoured.

incident radiation and so the additional sample is only absorbing 25% of the original incident radiation (Fig. 1.4). Doubling concentration has the same effect and the argument is equivalent.

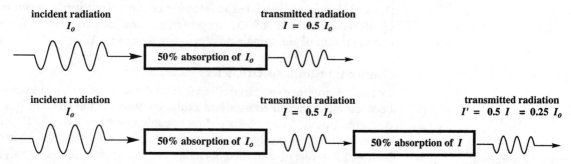

**Fig. 1.4**   Schematic illustration of effect of doubling path length upon absorption intensity

However the situation above only applies if every species capable of carrying out the absorption actually does so—in other words the solution is sufficiently dilute that no molecule is in the shadow of another. Under these circumstances the relationship between concentration, path length of the sample and the absorption is given by the **Beer–Lambert Law**:

$$\log\left(\frac{I_o}{I}\right) = \varepsilon c l$$

$I$ = transmitted radiation
$I_o$ = incident radiation
$\varepsilon$ = extinction coefficient
$c$ = sample concentration
$l$ = path length of the sample

The new term in the above expression, the **extinction coefficient** ($\varepsilon$) is a numerical reflection of the transitional probability and is constant for any given transition. If the transition is allowed, the extinction coefficient for the absorption will be large and *vice versa*. We will come across the Beer–Lambert expression in the analysis of ultraviolet spectra when sample strengths are so dilute that the above absorption condition is effectively met.

## 1.3   The spectrometer and data acquisition

In order to help with the understanding of absorption spectroscopy, it is useful at this stage to take a general overview of the instrumentation used to obtain the analyses. In absorption spectroscopy all spectrometers share several common features and consist of the following components:

1. The **radiation source**
2. The **monochromator** (continuous recording instruments)
3. The **sample container**
4. The **frequency analyser**
5. The **detector** → spectrally sensitive
6. The **recorder**

The specific construction of the instrument depends upon the electromagnetic radiation being used and the compound being analysed and a more detailed description of each spectrometer will be given at the introduction to each technique.

## Continuous recording spectrometers

The simplest way of recording absorption spectra, both conceptually and in the construction of the instrument, involves irradiating the sample sequentially through the wavelength range and recording the difference in intensities between the incident radiation, or a reference beam, and that passing through the sample. This output is charted directly on calibrated paper. The position of absorption gives information about the nature of the absorbing species while the intensity of the absorption is dependent upon both the ease of the transition to the excited state and the concentration of the sample as previously discussed.

### Fourier transform spectrometers

Instead of progressively irradiating and analysing differences in intensity between incident and transmitted radiation through the spectral region (the continuous mode of recording spectra described above), Fourier transform analysis involves irradiating the sample with the whole spectral bandwidth at one go and then detecting all frequencies simultaneously as a complex interference pattern. In FT–IR the resulting *interferogram* corresponds to the frequencies that pass through the sample and are not absorbed; whereas in FT–NMR the detected signal, referred to as the *free induction decay*, arises from excited species re-emitting the absorbed frequencies. In either case, digital sampling and submitting the data to Fourier transform analysis permits all the absorption frequencies to be extracted afterwards by computer.

As a macroscopic analogy to illustrate the difference between continuous scanning and Fourier transform modes of spectroscopic analysis, it is useful to consider two methods for obtaining the resonant frequencies of a bell. One method would be to use a range of tuning forks, each having different frequencies, apply them, one at a time to the surface of the bell and note which ones cause the bell to resonate. Alternatively, the bell could be struck, exciting all the resonant frequencies at once, and the resulting ringing emitted by the bell recorded and submitted to computer aided Fourier transform analysis. The latter approach is much faster than the continuous sampling method for acquiring data, but it leaves the problem of analysis until later.

Exciting all the resonant frequencies of a molecule and acquiring the emission spectrum leads to massive time saving compared to the continuous recording modes and means that multiple scanning is practical, giving spectacular improvements in signal-to-noise ratios and spectrometer sensitivity. This has revolutionized NMR spectroscopy, especially $^{13}C$ NMR analysis where the sampling is several thousands of times faster.

To record and interpret spectra, it is not necessary to understand in detail the means by which the complex signal of emitted frequencies is converted into a recognizable spectrum; but we do need to understand the relationship between the two sets of data. Mathematical manipulation of the interference pattern of various superimposed sine and cosine frequencies displayed in the *time domain* by the Fourier transform extracts the components and gives a *frequency domain* spectrum of the relative intensity of each signal plotted against its frequency (Fig. 1.5).

Consider sine waves oscillating at frequencies of 50 Hz, 100 Hz and 200 Hz respectively. Carrying out Fourier transform on each of these time domain oscillations gives frequency domain spectra, each of which

In Fourier transform NMR spectrometers, the sample is subjected to a broad pulse of radiation covering the whole of the frequency range. The absorbed frequencies which are re-emitted as the excited molecules decay back to their ground states are recorded together and the complex signal broken down to its component frequencies by subsequent computer assisted manipulation.

The Fourier transform approach to spectroscopic analysis improves spectrometer sensitivity and is sometimes referred to as *Fellget's advantage*.

Fourier transform analysis of the interference signal converts the time domain spectrum to the frequency domain spectrum.

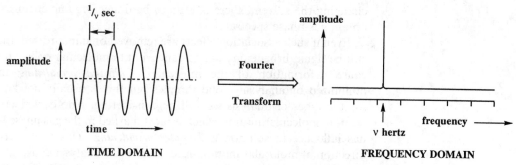

**Fig. 1.5**   Fourier transforming a time domain signal of a single frequency into a frequency domain signal.

consists of a single peak occurring at the frequency of the sine wave (Fig. 1.6a–c). Progressively combining the time domain frequencies gives an increasingly complex interference pattern which nevertheless can be subjected to Fourier transform analysis to extract all the component frequencies (Fig. 1.7a, b). However, these examples do not represent what actually happens during spectroscopic analysis as they have constant amplitude. In NMR, the signal emitted by the excited spins decreases in amplitude as the nuclei relax back to their ground state (Fig. 1.7c). Nevertheless, despite the fact that the interference pattern dies away, it can still be transformed to obtain the frequency domain signal. A fuller consideration of this technique will be given in Chapter 5.

## Resolution

An important feature of any spectrometer's performance is its **resolving power**; in other words the ability of the instrument to distinguish between two particular frequencies or wavelengths of absorption. No molecular absorption of electromagnetic radiation takes place at a single frequency, so there will always be a spread of frequencies associated with any specific absorption process. This spread may be very narrow, such as in the case of gas-phase ultraviolet spectra of simple molecules and NMR signals, or it may be particularly broad, such as in solution phase ultraviolet spectra or infrared spectra. The degree of broadening depends upon various factors, some of which are a consequence of the molecular environment of the species under examination and are largely not under our control. These give rise to the **natural linewidth** which is a degree of resolution which no instrument, no matter how precise, can surpass. However, some effects are due to the construction and mode of operation of the spectrometer and this **resolving power** may be amenable to modification.

   Line–broadening effects largely beyond our control have their origin in the fact that analysis is carried out on the bulk sample which consists of a range of slightly different absorbing species, no matter how much care is taken to ensure sample homogeneity. For example, intermolecular associations, notably by ionic, dipole or hydrogen bonding interactions, result in mixtures of monomers, dimers and higher oligomers. Associations of this nature can be controlled to some extent by diluting the sample. This will decrease the degree of self-association, although a price has to be paid in decreasing signal strength. This does not remove solvation effects however and, apart from attempting to modify these by

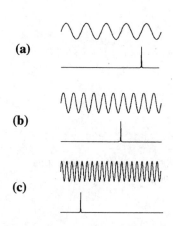

**Fig. 1.6**   Effects of carrying out Fourier transforms on sine waves having frequencies (a) 50 Hz (b) 100 Hz (c) 200 Hz.

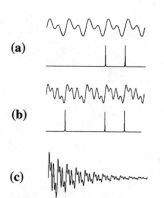

**Fig. 1.7**   Effects of carrying out Fourier transforms on progressively more complex interference patterns. (a) 50 + 100 Hz (b) 50 + 100 + 200 Hz (c) 50 + 100 + 200 Hz FID.

changing the solvent, there is little to be done. Organic chemists rarely obtain gas-phase spectra.

Even if such association effects are removed or standardized, there will still be signal broadening due to random intermolecular collisions which cause deformation of the molecules. **Collision broadening** may be minimized by analysing solid samples but this is usually not an option for the types of analysis we will be considering. Another effect due to random molecular motion which is most marked in the gas phase but also has influence in solution is **Doppler broadening**. This results from the direction of molecular motion in relation to the analyser causing shifts to slightly higher or lower frequencies of the true absorption frequency.

Even if it were possible to analyse an individual molecule, it is impossible to isolate a specific transition totally from other molecular transitions. For example, the vibrational frequency of a carbonyl group in a molecule is to some small extent modulated by other vibrational frequencies within the molecule as a whole.

Finally, even with a stationary isolated absorbing species, energy levels are not perfectly defined, as **Heisenberg's uncertainty principle** tells us that a system remaining in a fixed energy for a time $\Delta t$ sec will have a degree of uncertainty of the energy of that state which is defined by the following equation:

$$\Delta E \times \Delta t \approx \frac{h}{2\pi} \approx 10^{-34}\,\text{Js}$$

$\Delta E$ = degree of uncertainty of the energy level.

Note that the ground state energy of a system *is* sharply defined as, in the absence of external influences, the system would stay there for infinite time and therefore $\Delta E = 0$. However, all other energy states have uncertainty relating to their lifetime and consequently there will be an uncertainty in the transition energy, and hence frequency of absorption of radiation, which is described by:

$$\Delta v = \frac{\Delta E}{h} \approx \frac{1}{2\pi\Delta t}$$

It turns out that, despite the short lifetimes of electronically and vibrationally excited species the degree of uncertainty, $\Delta v$, is small compared to the frequency of the transitions. However, with electron and nuclear spin transitions, the small energy differences between states give rise to uncertainties which are more significant in relation to the lower frequency radiation causing the transitions.

If all of the above effects contribute to a certain blurring of the absorption which we can do little to affect, there are nonetheless some features relating to the spectrometer over which we do have some control.

In continuous scanning ultraviolet and infrared spectrometers the major controlling elements on the resolving power are the quality of the optical arrangement and the **slit width**. Using diffraction gratings, much better dispersion of the radiation is possible compared with prism based optics and the limiting factor is usually the slit width of the spectrometer. The slit width refers to the size of the sampling window of the spectrometer. Obviously, the narrower the slit width, the higher the resolving power of the instrument and this is illustrated in Fig. 1.8.

However, if the slit becomes too narrow, insufficient energy will reach the detector and the signal-to-noise ratio becomes unacceptable. Using

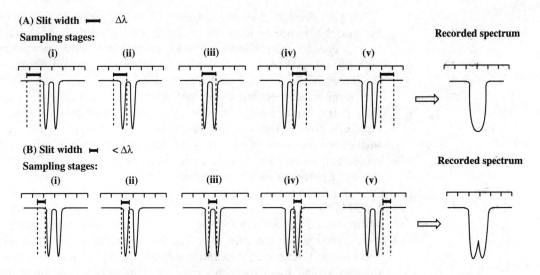

**Fig. 1.8** Dependence of resolution upon slit width in continuous recording IR and UV spectrometers.

more sensitive detectors permits narrower slit widths and hence better resolving power of the spectrometer but always a compromise must be reached.

In Fourier transform instruments, the smaller the difference between two frequencies, the longer it is necessary to record the free induction decay in order to resolve the component frequencies. As an approximation, a resolution of $\Delta v$ requires the emission signal to be sampled every $\frac{1}{\Delta v}$ seconds and this period is known as the *acquisition time*. However, there is a limit to this time as, once the emission has decayed, no further information can be acquired (Fig. 1.9).

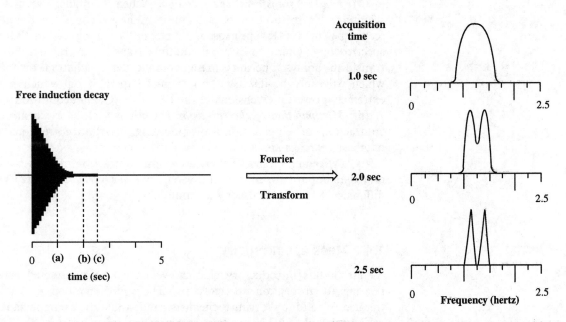

**Fig. 1.9** Schematic showing the dependence of resolution upon acquisition time in FT spectrometers. Note that attempting to acquire data after point (c) will not lead to greater resolution as the emission signal has decayed to zero at this stage.

This requirement does not cause problems with infrared spectrometers because the resolution required means that a few milliseconds suffice for the acquisition time However, NMR spectra are commonly recorded with a resolution of 0.3 to 0.5 Hertz which requires acquisition times in the order of 2 to 3 seconds. Once again, it is necessary to strike a compromise between resolution and sensitivity, because longer acquisition times mean fewer scans in any given time period. Evidently, acquiring for longer than the decay time of the FID cannot lead to further enhancement of resolution as there is no more emission signal to record and so the rate of decay of the signal places a limiting value on the resolving power of the spectrometer.

### Computation

With the advent of cheap but powerful computing capability, new instrumentation techniques have become available which speed up the processes of data acquisition and manipulation, and also make it possible to obtain high quality spectra from progressively smaller amounts of material (particularly important for NMR spectra). Computers render possible the following processes:

**Noise** is the term applied to random signals recorded by the spectrometer.

(i) *Spectrum summation*. This is used for analysing weak samples to improve signal-to-noise ratio. Noisy spectra can be improved by adding together the data from repeated analyses of the same sample. Noise, being random, will tend to average out; whereas the peaks will always be additive. It turns out that the intensity of a peak increases at a rate proportional to the square root of the number of spectra recorded. In other words, it is necessary to record 100 spectra of the same sample to obtain a ten-fold improvement in signal-to-noise. This is often known as *CAT scanning* (Computer Averaging of Transients) and is most time efficient and now routine in the FT mode.

(ii) *Fourier transform spectroscopy*. When combined with CAT scanning this method of data acquisition has become the dominant technique in NMR spectroscopy and is becoming so in infrared spectroscopy. Quite apart from multiple transient averaging, Fourier transform analysis permits many complicated irradiation protocols, which will only be discussed briefly in Chapter 5, but which permit extremely powerful extensions of the basic NMR data acquisition.

(iii) *Manipulation of the spectrum*. Functions such as expansion and contraction of abscissas, noise smoothing, baseline corrections and database searches are commonly available facilities.

(iv) *Difference spectra*. This is useful for subtraction of known peaks due to an impurity or solvent and also in *nuclear Overhauser effect* (*NOE*) difference NMR spectroscopy (Section 5.3).

### 1.4   Mass spectrometry

So far in this introductory chapter we have only considered aspects relating to absorption spectroscopy. The fourth method of analysis frequently used by organic chemists, and arguably as important an analytical tool as NMR, is *mass spectrometry*.

As its name suggests, this technique allows measurement of the relative molecular mass of molecules and therefore provides vital information not available by spectroscopic methods. Quite apart from the fact that this technique does not involve absorption of electromagnetic radiation (note the terminological distinction of spectro*metry versus* spectro*scopy*), it differs from the other three techniques by analysing sample molecules one at a time rather than the bulk sample.

Mass spectrometry enables the molecular weight of quite large and complicated molecules to be obtained. In addition, the fragmentation of the excited species produced by ionization of the subject molecule, frequently gives very important clues to its composition. Therefore, mass spectrometry complements absorption spectroscopy by providing us with important information about the mass of the intact molecule and additional details about its construction. We will consider this technique in more detail in Chapter 6.

## 1.5　Summary

In this chapter we have been introduced to the four types of analytical technique most frequently used by organic chemists in structural analysis. In addition, we have considered in a very qualitative way the physical basis behind absorption spectroscopy and some of the principles behind the instrumentation. We will now proceed to consider each of the four techniques in more detail. Each chapter will start with a qualitative treatment of the theoretical principles and and then go on to analysis of the data. At the end of each chapter will be a series of exercises to help assess your understanding of the topics covered and there will be lists of additional sources of information for more detailed or theoretical coverage.

In this introduction to organic spectroscopy it has been necessary to make somewhat arbitrary decisions about which techniques will be treated and which will be left out. Three types of absorption spectroscopy (ultraviolet, infrared and NMR spectroscopy), together with mass spectrometry, normally provide the organic chemist with sufficient analytical power to determine the structure of even the most complex organic molecule and will be the only techniques considered. Furthermore, the sections on NMR spectroscopy will be limited to $^{1}$H and $^{13}$C NMR spectroscopy, although nuclei such as $^{2}$H, $^{19}$F and $^{31}$P may also be of interest to organic chemists. Likewise, we will not consider electron spin resonance spectroscopy, a means of detecting species which possess an unpaired electron, nor will we discuss chiroptical techniques such as circular dichroism and optical rotary dispersion. However, we must not forget that such techniques are available, and at times necessary, and those wishing to have a simple introduction to any of the areas not covered by this book are directed to the *Further reading* section at the end of this chapter.

Equally, the confines of this introductory book mean that it has not been possible to include a problem section involving structural elucidation using the range of spectroscopic techniques and you are recommended to consult the texts in reference 4. These contain a series of exercises which should be within your scope after having read this book.

## 1.6    Exercises

1.    Calculate the frequency of infrared light of wavelength $\lambda = 15$ μm. ($1 \, \mu m = 10^{-6}$m, $c = 3 \times 10^8$ ms$^{-1}$)

2.    In infrared spectroscopy it is customary to express absorption positions as *wavenumbers* (wavenumber = reciprocal of wavelength and units are cm$^{-1}$). What is the wavenumber of the above frequency?

3.    Calculate the frequencies and wavelengths of electromagnetic radiation having wavenumbers of 1750 cm$^{-1}$ and 3500 cm$^{-1}$.

4.    For UV light of wavelength 254 nm calculate the frequency and the amount of energy absorbed by a single molecule when it absorbs a quantum of this light. What is the corresponding energy (in J) absorbed by one mole of a molecule absorbing at this frequency ($h = 6.6 \times 10^{-34}$ Js, Avogadro's number $N_A = 6.02 \times 10^{23}$ mol$^{-1}$)?

5.    What frequency and wavelength of electromagnetic radiation is necessary to cause rupture of a C–C bond with a dissociation energy of 380 kJ mol$^{-1}$? To what region of the electromagnetic spectrum does this correspond?

## Further reading

1.    For further introductory texts to spectroscopy with the emphasis on non-mathematical treatment see: C. N. Banwell and E. M. McCash, *Fundamentals of Molecular Spectroscopy*, 4th edn, 1994, McGraw-Hill, London; W. Kemp, *Organic Spectroscopy*, 3rd edn, 1991, McMillan, Basingstoke.

2.    For texts covering specific areas not treated in this introduction see: D. P. Strommen and K. Nakamoto, *Laboratory Raman Spectroscopy*, 1984, Wiley Interscience, New York; P. B. Ayscough, *Electron Spin Resonances in Chemistry*, 1967, Methuen; C. Djerassi, *Optical Rotatory Dispersion*, 1960, McGraw-Hill; P. F. Ciardelli and P. Salvadori (Eds), *Fundamental Aspects and Recent Developments in Optical Rotatory Dispersion and Circular Dichroism*, 1973, Heyden and Son Ltd, London; N. Harada and K. Nakanishi, *Circular Dichroic Spectroscopy—Exciton Coupling in Organic Stereochemistry*, 1983, Oxford University Press, New York. A text which covers the whole range of spectrometric techniques is: J. D. Ingle and S. R. Crouch, *Spectrochemical Analysis*, 1988, Prentice Hall, New Jersey.

3.    For a detailed yet still relatively qualitative survey of the applications of Fourier transform techniques in NMR spectroscopy see: A. E. Derome, *Modern NMR Techniques for Chemistry Research*, 1987, Pergamon, Oxford. See also: R. P. Wayne, *Fourier Transformed*, in *Chemistry in Britain*, 1987, **23**, 440.

4.    For structural elucidation exercises using a combination of UV, IR, NMR and MS see: L. D. Field, S. Sternhell and J. R. Kalman, *Organic Structures from Spectra* 2nd edn, 1995. John Wiley and Sons Ltd, Chichester: R. M. Silverstein, G. C. Bassler and T. C. Morrill, *Spectrometric Identification of Organic Compounds*, 5th edn, 1991, John Wiley and Sons Ltd, New York.

# 2 Ultraviolet–visible spectroscopy

In a historical context, the introduction of UV spectroscopic analysis during the 1930s was an important milestone in analytical chemistry as it demonstrated the advantages of non-destructive analysis. The vast savings in time and the ability to work with much smaller samples of material demonstrated the power of spectroscopic analysis to organic chemists who wanted to use the technique in a diagnostic manner without being concerned with the full details of the physical background. This paved the way for the development and acceptance of the other techniques. It is worthy of note that, powerful as it seemed at the time, UV spectroscopy is rarely used as a first means of analysis nowadays. Its use is commonly restricted to very specific areas of chemistry dealing with compounds, such as dyestuffs, which have important absorption characteristics in this range of the electromagnetic spectrum. Despite this, it is worthwhile considering UV–visible spectroscopy first as it provides us with a simple introduction to the major principles of absorption spectroscopy.

## 2.1 Absorption of UV–vis electromagnetic radiation

Absorption of electromagnetic radiation in the UV–vis region transfers the right amount of energy to cause transitions in the electronic energy levels of the bonds of a molecule and results in excitation of electrons from the *ground state* into an *excited state* (Fig. 2.1). However, as discussed in Chapter 1, absorption will only occur if the energy of the photon corresponds exactly to that of the transition. We will see that higher energy, shorter wavelength, light is necessary to excite σ-bonds than π-bonds as there is a larger energy difference between the *bonding* and *antibonding* σ-orbitals compared with the bonding and antibonding π-orbitals. In spectroscopic nomenclature the structural features of a molecule responsible for absorption of UV light are referred to as *chromophores*.

Molecules absorb UV radiation in the region below about 150 nm as these wavelengths of electromagnetic radiation have energies corresponding to those required to cause excitation of electrons in the σ-bond framework. Consequently molecules in the atmosphere will absorb this radiation, and therefore analysis in this region requires special vacuum techniques.

In standard UV–vis spectroscopy analysis is carried out on a dilute solution of the sample and referenced to a sample of pure solvent. In this situation, with the solvent present in a vast excess, the tail of the solvent σ-bond absorption peak will impinge into the measurable range of the UV spectrometer and totally swamp any absorptions due to the dissolved compound under investigation, as the spectrometer becomes insensitive when attempting to subtract two extremely strong absorptions. *Solvent*

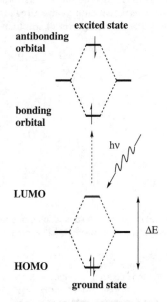

**Fig. 2.1** Absorption of a photon having exactly the same energy as the difference between the *bonding* and *antibonding* orbitals can lead to the excitation of one electron between these levels.

***end absorption*** may extend up to about 240 nm, depending upon the solvent. Certainly any peaks observed below 220 nm must be regarded with caution. Some common solvents and their transparency ranges are given below.

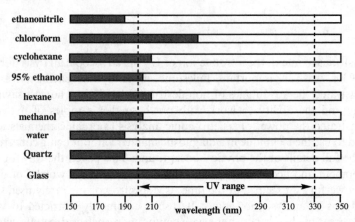

**Fig. 2.2**   Useful transparency ranges in the region 180–250 nm (▭) for common solvents in UV–visible spectroscopy.

In addition to considering solvent, it must be remembered that ordinary glass is opaque to wavelengths below 300 nm. UV analysis is carried out in special quartz vessels (***cuvettes***) which are transparent down to about 170 nm, but visible spectra may be obtained using glass or plastic cuvettes.

> The ultraviolet range is generally considered to be 200–330 nm and the visible range 330–700 nm. Of course, if the sample is not coloured, there is no need to record the visible region of the spectrum.

Electrons in π-bonds and lone pairs require less energy than σ-bonding electrons for excitation to their antibonding orbitals and absorb at longer wavelengths. An isolated double bond absorbs UV-radiation at about 190 nm and lone pairs are excited by wavelengths between 185–195 nm. The table below lists the important fundamental chromophores and the types of electronic transition they undergo. With the exception of the ***symmetry forbidden*** n → π* transition, none occur within the recording range of standard instruments.

However, when two π-orbitals are combined to form a ***conjugated*** diene

> σ* and π* refer to the antibonding σ- and π-bonds. n refers to non-bonding orbital.

> Although occurring at longer wavelength than π → π* and σ → σ* transitions, the n → π* transition is said to be ***symmetry forbidden*** and is very weak.

| Chromophore | | Electronic Transition | $\lambda_{max}$ nm |
|---|---|---|---|
| σ–bonded systems | C–C, C–H | σ -> σ* | ~ 150 |
| lone pairs | –Ö– | n -> σ* | ~ 185 |
| | –N< | n -> σ* | ~ 195 |
| | –S– | n -> σ* | ~ 195 |
| | C=Ö | n -> σ* | ~ 190 |
| | C=Ö | n -> π* | ~ 300 |
| π–bonded systems | C=C | π -> π* | ~ 190 |

**Fig. 2.3**   Chromophores and their related electronic transitions.

the $\pi$-electrons become delocalized over the whole $\pi$-framework giving rise to a resonance hybrid structure. The energy gap between the ***highest occupied molecular orbital (HOMO)*** of this resonance hybrid and the ***lowest unoccupied molecular orbital (LUMO)*** is smaller than in the isolated alkene. Consequently, excitation occurs with a longer wavelength of electromagnetic radiation. Considering Fig. 2.4 showing the molecular orbitals for an alkene and a conjugated diene allows us to see how conjugation leads to a lowering of the HOMO–LUMO gap compared with the isolated system.

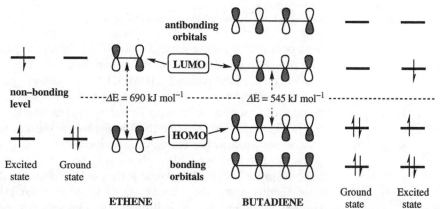

**Fig. 2.4**   The molecular orbitals of ethene and butadiene. Note how conjugation reduces the HOMO–LUMO energy difference.

Inclusion of additional $\pi$-bonds and extending the degree of conjugation progressively narrows the gap between the energy of the HOMO and the LUMO. As a result, the wavelength of radiation required for excitation becomes longer as the degree of conjugation increases. It is conjugated systems such as polyenes and poly-ynes, referred to as ***chromophores***, that give rise to diagnostic absorptions in the UV–visible region of the electromagnetic spectrum.

The term ***chromophore*** originates from the fact that conjugation was recognized to be responsible for making compounds coloured (the absorption reached into the visible region of the spectrum) but it now applies to any conjugated system resulting in absorption of UV or visible light.

The diagnostic absorptions result from $\pi \rightarrow \pi^*$ and $n \rightarrow \pi^*$ transitions in conjugated systems. Aromatic and heteroaromatic systems are commonly encountered examples of conjugated $\pi$-systems and such structures are particularly useful chromophores. It is with such conjugated systems that UV–visible spectroscopy comes into its own.

The non-bonding electrons of heteroatoms such as nitrogen, oxygen and sulphur can also become involved in resonance and will extend conjugation in these systems. An ***auxochrome*** is any group, such as OR, $NR_2$, $NO_2$, or $CO_2R$, which is capable of extending the basic chromophore. This synergistic effect is due to the ability of such substituents to undergo resonance delocalization involving the polyene system and so reduce the HOMO–LUMO energy gap still further.

Representative spectra shown in Fig. 2.5 demonstrate how increasing conjugation in a series of related structures gives rise to increase in the wavelength of maximum absorption ($\lambda_{max}$).

Although absorptions in UV–vis solution spectra are broad, the ***wavelength at maximum absorption*** ($\lambda_{max}$) is quoted for each absorption.

As well as defining the lower wavelength limit of analysis, the solvent in which the UV spectrum is measured can have an effect on the position of $\lambda_{max}$. The ***Franck–Condon principle*** states that, during electronic excitation, atoms do not move, but electrons—including those of the solvent—may reorganize themselves. Most transitions lead to excited states which are

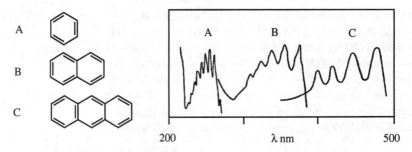

A

B

C

**Fig. 2.5**   Increasing conjugation results in an increase in the wavelength of maximum absorption ($\lambda_{max}$).

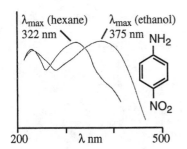

**Fig. 2.6**   The bathocromic effect in the UV spectra of 4-nitroaniline recorded in hexane and ethanol.

more polarized than the ground state. If the solvent is polar, the excited state can be more readily stabilized by dipole–dipole interactions than the ground state. The reduced HOMO–LUMO energy gap means that excitation occurs more readily than in apolar solvents, resulting in a 10–50 nm shift to longer wavelength (lower energy) in ethanol compared with hexane. Such shifts are referred to as **bathochromic shifts** or **red shifts** as the $\lambda_{max}$ moves towards the red end of the spectrum.

Exceptions are those absorptions due to the $n \rightarrow \pi^*$ transition of carbonyl groups when the ground state can hydrogen-bond with solvents more readily than the excited state. Electronic excitation therefore requires more energy in hydrogen-bonding solvents compared with apolar solvents and, in these instances, the wavelength of light absorbed shows a shift of about 15 nm to shorter wavelength. This **hypsochromic shift** or **blue shift** is a consequence of the transition requiring higher energy electromagnetic radiation in a more polar solvent.

## 2.2   Recording and interpreting UV–vis spectra

UV spectroscopy gives two important pieces of information. One is the position of absorption ($\lambda_{max}$ nm) which we have already encountered. This gives information on the energy ($\Delta E$) of the electronic excitation taking place and, as a result of a great number of empirical correlations, information on the actual chromophore responsible for the absorption.

The second value is the **molar absorptivity**, also known as the **molar extinction coefficient** ($\varepsilon$). This value, which is a constant for a given molecule at a given wavelength, is a measure of the ease of the transition caused by the absorption of the radiation. If a particular electronic transition requiring a certain energy of incident radiation can take place readily then the light will be absorbed strongly and the $\varepsilon$ value will be high. If the transition does not occur readily then the $\varepsilon$ value will be low. Such electronic transitions are often distinguished as **allowed** and **forbidden** transitions (although for forbidden read 'strongly disfavoured') and may be predicted on the grounds of symmetry correlations of the ground state and excited state of the molecule. For example, $\sigma \rightarrow \sigma^*$ and $\pi \rightarrow \pi^*$ transitions are 'allowed'; whereas $n \rightarrow \pi^*$ transitions are 'forbidden'. However, for analysing UV–vis spectra we simply need to be aware that allowed processes have $\varepsilon$ values of greater than $10^3$ and forbidden processes have $\varepsilon$ less than 100. Substituents which increase the extinction coefficient of the basic chromophore are said to have a

***hyperchromic effect*** whereas those decreasing the extinction coefficient have a ***hypochromic effect*** (p. 19).

Of course, comparison between ε values for different compounds is only valid if the analyses are run with solutions at the same molarity, increasing the amount of sample present will naturally increase the ability of the solution to absorb incident radiation. The relationship between the molar absorptivity, ε, and the concentration of chromophore (in mol $L^{-1}$) is expressed by the ***Beer–Lambert Law***:

$$\log\left(\frac{I_0}{I}\right) = \varepsilon c l$$

The equation is often stated as:

$$\varepsilon = A/cl$$

Note that the formula results in the molar absorptivity having the rather absurd units of $cm^2 \, mol^{-1}$ but, by convention, these are never stated.

UV–visible spectrometers record absorbance directly and the standard cuvette is designed to give a path length of 1 cm. Thus determination of ε is a straightforward exercise as the equation becomes:

$$\varepsilon = A/c$$

The derivation of this relationship assumes that all molecules contribute to absorption of the incident beam and that no absorbing molecule is in the shadow of another. For the sorts of very dilute concentrations used in UV–vis spectroscopy this is a valid assumption (Fig. 2.7).

### The instrument

UV–visible spectrometers operate on the 'double beam' principle, with one beam passing through the sample and the other passing through a reference cell (Fig. 2.8). The spectrometer has two lamps, one emitting light in the range 200–330 nm (UV) and the other in the range 330–700 nm (visible). The light emitted by the source first passes through a diffraction grating ***monochromator***, which breaks down the light into its component wavelengths in the same way as a prism. This light is then divided by curved mirror optics into two beams of equal intensity, one of which passes through the sample, the other through the reference

**Hyperchromic** substituents increase the extinction coefficient.
**Hypochromic** substituents decrease the extinction coefficient.

$I_0$ = Intensity of incident light
$I$ = intensity of transmitted light
$\varepsilon$ = molar absorptivity
$c$ = concentration in mol $L^{-1}$
$l$ = path length of absorbing solution in cm.
$A$ = **absorbance** = $\log(I_0/I)$.

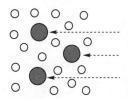

(a) dilute solution

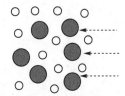

(b) concentrated solution

**Fig. 2.7** UV–vis analysis concentrations meet the conditions for the Beer–Lambert law to be valid.

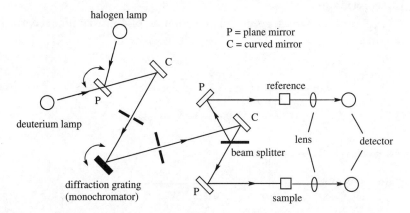

**Fig. 2.8** Optical arrangement of a double beam UV–visible spectrometer (Adapted and reproduced with permission of Perkin Elmer).

containing pure solvent. Each beam is then focused onto a ***detector*** which measures the ratio of the intensities of the two beams and the difference is automatically converted and plotted out as the absorbance. Instruments usually have a full scale deflection for $A = 2$, which means that the strongest ratio of $I_0/I$ that the instrument can tolerate is 100.

### Recording the spectrum

The UV spectrum of a compound is recorded as a dilute solution in a $1 \text{ cm}^2$ cross-section quartz cuvette (glass is opaque to UV below about 300 nm). The sample is placed in the analysis beam path and a matched cuvette containing pure solvent is placed in the reference beam path. The spectrometer records the difference in intensities of the radiation passing through each of the cuvettes in order to give a measure of the light absorbed by the dissolved substance under analysis. The substance must be soluble in the solvent, but thought must be given to any possible solvent shifts and 'cut off' due to solvent end absorption (Section 2.1). Frequently it may be necessary to record spectra over a range of different concentrations in order to bring various peak intensities within the measurable range of the spectrometer.

### Appearance of the spectrum

Due to the presence of variously solvated and associated species, UV–visible spectra recorded in solution are generally poorly resolved and consist of one or more broad humps, sometimes with some poorly defined fine structure. Often peaks of different intensities will coalesce so that the smaller peak appears as a shoulder on the larger one.

Whilst in theory it should be possible to cancel out the solvent absorption, the detector cannot handle small differences in very large values and so erratic results tend to be observed at the 200–210 region which can be mistaken for peaks by the unwary. These spurious peaks can be easily detected as they do not diminish in height on diluting the sample and re-recording the spectrum; whereas genuine peaks will become progressively smaller. A discontinuity may be noticed in the trace at 330 nm when the radiation source switches from the UV lamp to the visible lamp and may be ignored.

### Interpretation

Fig. 2.9 shows a typical spectrum of a solution of 4-nitroaniline recorded in ethanol.

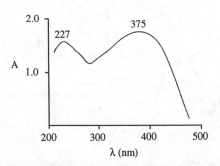

**Fig. 2.9**  A typical UV–vis spectrum of 4-nitroaniline in EtOH recorded 15.4 mg L$^{-1}$ (obtained by successive ten-fold dilutions of a stronger solution), path length = 1 cm.

Firstly, determine the $\lambda_{max}$ and $\varepsilon$ value(s) for the spectrum.

$\lambda_{max}$ 227 nm, $A = 1.55$; $\lambda_{max}$ 375 nm, $A = 1.75$

molecular weight of 4-nitroaniline = 138.
Molarity of solution = $1.12 \times 10^{-4}$ mol L$^{-1}$

$$\varepsilon = A/cl$$

$\varepsilon_{227} = 1.55/1.12 \times 10^{-4} = 13\ 890$

$\varepsilon_{375} = 1.75/1.12 \times 10^{-4} = 15\ 680$

Quote these results as $\lambda_{max}$ (EtOH) 227 ($\varepsilon$ 13 890), 375 ($\varepsilon$ 15 680).

## Benzene Derivatives

Benzene shows three absorption maxima in hexane at 184 ($\varepsilon$ 60,000), 203.5 ($\varepsilon$ 7400) and 254 ($\varepsilon$ 204) nm (Fig. 2.10). The major peak is termed the **K-band**. The latter, forbidden absorption is called the **B-band** and shows fine structure due to excitation of vibrational energy levels due to 'ring breathing' vibrations of the benzene ring; although the so called $0 \rightarrow 0$ transition (the transition between the vibrational ground state levels of the electronic ground state and the excited state) is not observed within this set of peaks. However, substitution of the ring or the presence of a heteroatom in the ring, such as in pyridine, lowers the symmetry further and the $0 \rightarrow 0$ transition is then observed.

Substituting the benzene ring with conjugated substituents (e.g. $CO_2H$ or $CH = CHCO_2H$) or electron donating groups (e.g. $NR_2$, $OR$) extends the chromophore and results in a bathochromic shift of the peaks although generally not in a predictive manner.

Two special situations worthy of note occur with hydroxybenzenes (phenols) and aminobenzenes (anilines) as these show very characteristic behaviours on treatment with base and acid respectively.

Adding base to a phenol causes a marked bathochromic (red) shift of the absorption maximum as the lone pair on the oxygen becomes more available for delocalization into the aromatic nucleus (Fig. 2.11a).

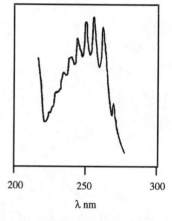

**Fig. 2.10** The UV spectrum of benzene recorded in cyclohexane showing the vibrational fine structure of the B-band.

**(a)**

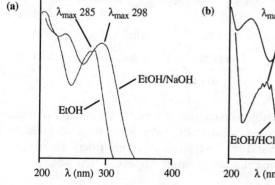

$\lambda_{max}$ 285    $\lambda_{max}$ 298

EtOH/NaOH

EtOH

200    $\lambda$ (nm) 300    400

**(b)**

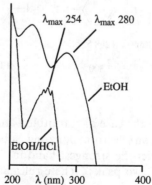

$\lambda_{max}$ 254    $\lambda_{max}$ 280

EtOH

EtOH/HCl

200    $\lambda$ (nm) 300    400

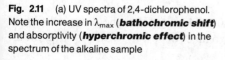

**Fig. 2.11** (a) UV spectra of 2,4-dichlorophenol. Note the increase in $\lambda_{max}$ (**bathochromic shift**) and absorptivity (**hyperchromic effect**) in the spectrum of the alkaline sample

(b) UV spectra of aniline. Note the decrease in $\lambda_{max}$ (**hypsochromic shift**) and absorptivity (**hypochromic effect**) in the spectrum of the acidic sample which shows a marked similarity to that of benzene (Fig. 2.10).

Conversely anilines show a marked hypsochromic (blue) shift of the absorption maximum upon acidification as the lone pair is no longer available for delocalization in the protonated anilinium species (Fig. 2.11b).

When electronically 'complementary' substituents (in other words, when one group is mesomerically electron-donating, such as $NH_2$, OH, OR, and one group is mesomerically electron-accepting, such as $NO_2$, COR, $CO_2R$) are present on the ring in a 1,4-relationship to each other, a very pronounced bathochromic shift and hyperchromic effect is noted on the main absorption band (Fig. 2.12).

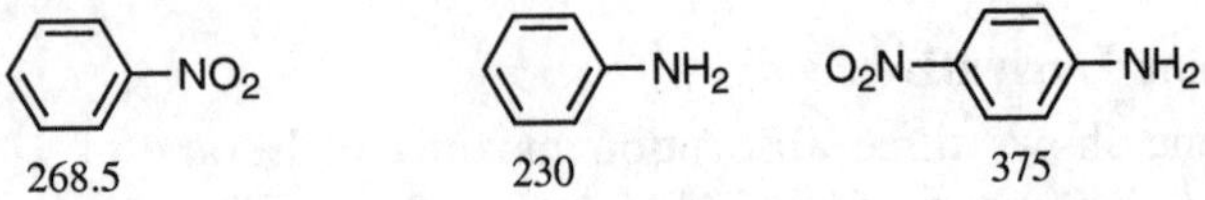

| | $NO_2$ | $NH_2$ | $O_2N$—❪—$NH_2$ |
|---|---|---|---|
| $\lambda_{max}$ (nm) | 268.5 | 230 | 375 |
| $\varepsilon$ | 7 800 | 8 600 | 16 000 |

**Fig 2.12**

This effect is due to extension of the chromophore by resonance delocalization across the aromatic ring. However, when complementary substituents are 1,2- or 1,3- to each other this bathochromic effect is not noticeable although the hyperchromic shift remains (Fig. 2.13).

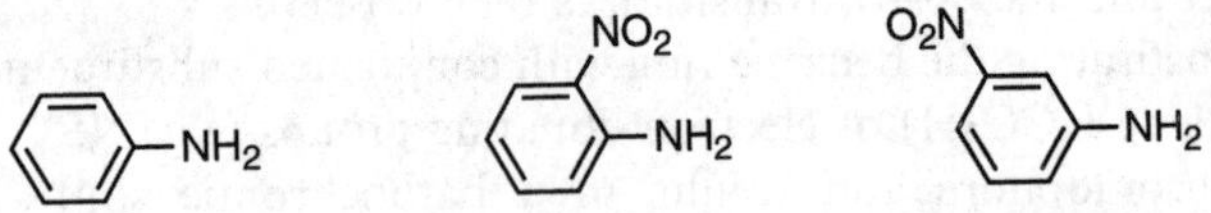

| | $NH_2$ | $NO_2$ / $NH_2$ | $O_2N$ / $NH_2$ |
|---|---|---|---|
| $\lambda_{max}$ (nm) | 230 | 229 | 235 |
| $\varepsilon$ | 8 600 | 14 800 | 16000 |

**Fig 2.13**

Similarly with non-complementary substituents there is generally little bathochromic effect even when the groups have a 1,4-relationship (Fig. 2.14).

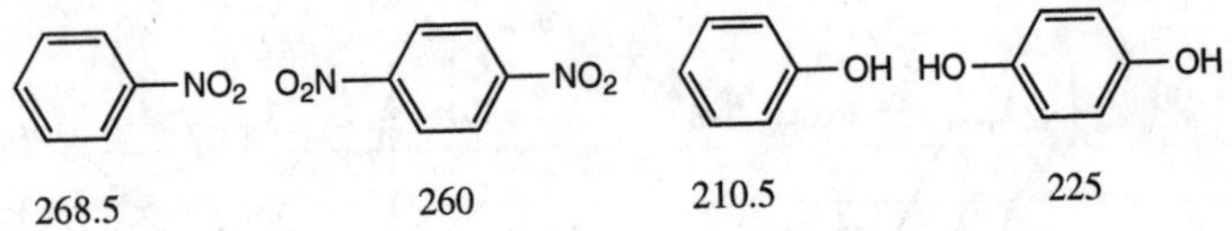

| | $NO_2$ | $O_2N$—❪—$NO_2$ | OH | HO—❪—OH |
|---|---|---|---|---|
| $\lambda_{max}$ (nm) | 268.5 | 260 | 210.5 | 225 |
| $\varepsilon$ | 7 800 | 13 000 | 6 200 | 5 100 |

**Fig 2.14**

In the case of substituted aromatics possessing a carbonyl substituent and an electron donating substituent, rules have been formulated to predict the strongest band in the UV–vis spectrum (there are usually at least two peaks). These rules are given in Appendix 1 (p. 88).

## 2.3  Exercises

1.  The UV spectrum of acetone (0.05M in cyclohexane, 1 cm cell) shows $\lambda_{max}$ at 279 nm with an absorbance of 0.74. What is the molar

absorptivity of acetone at this wavelength? Is it an allowed or forbidden transition? What is the transition which is causing this absorption?

2.  But-3-en-2-one shows UV absorption maxima at 213 and 320 nm. Why are there two maxima and which should have the greater $\varepsilon$ value?

3.  A liquid compound of molecular formula $C_5H_8$ has a $\lambda_{max}$ at 220 nm. Suggest possible structures for the compound.

4.  Predict $\lambda_{max}$ values for the following compounds using the tables of values given in Appendix 1.

5.  Using the tables in Appendix 1 match the steroid derivatives below with the following values for $\lambda_{max}$ in hexane: 240 nm, 289 nm, 355 nm.

6.  Calculate the molar absorptivity for the following compounds at the maximum wavelengths given recorded in 1 cm pathlength cells.

| | conc (mg L$^{-1}$) | $\lambda_{max}$ (A) | | |
|---|---|---|---|---|
| | 8.9 | 257 (1.05) | | |
| | 18.4 | 252 (1.65) | 282 (0.22) | 292 (0.14) |
| | 14.0 | 252 (1.50) | | |

## Further reading

1.  D. H. Williams and I. Fleming, *Spectroscopic Methods in Organic Chemistry*, 5th edn, 1995, Chapter 1, McGraw–Hill, London.

2.  W. Kemp, *Organic Spectroscopy*, 3rd edn, 1991, pp. 243–268 McMillan, Basingstoke.

3.  For important collections of reference data see: *Organic Electronic Spectral Data*, (Vols 1–21) Wiley, New York, 1960–1985; *Sadtler Handbook of Ultraviolet Spectra*, 1979, Sadtler, Pennsylvania.

# 3  Infrared spectroscopy

The principles behind infrared spectroscopy are the same as those already discussed for ultraviolet spectroscopy except that this time the energy range of the infrared region corresponds to that required to cause **vibrational excitation** of bonds within a molecule (we will assume much of what has already been covered and simply modify it as necessary for this technique). The types of bond excitation which can occur are **stretching** (higher energy) and **bending** (lower energy) vibrations (Fig. 3.1).

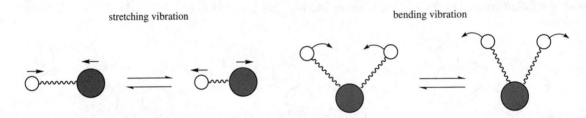

stretching vibration                                   bending vibration

**Fig. 3.1**  'Ball and spring' schematic diagram showing the two types of vibrational motion in molecules.

Often, absorption of certain wavelengths of infrared radiation can be correlated with bending or stretching of specific types of bond within a molecule. However, it must be remembered that any single vibrational excitation cannot be considered in isolation from other parts of the molecule. Infrared spectra of organic molecules are usually complicated by bond oscillations in the whole molecule affecting absorption of the incident radiation and giving rise to **overtones** and **harmonics**. Therefore, in addition to observing absorptions due to individual bond vibrational excitations, molecular vibrational excitations are also observed.

Spectra recorded in solution or as a neat substance—a liquid film for instance—may also show further complexity due to hydrogen bonding with solvents or the presence of dimeric or polymeric associated species. Gas phase spectra (like UV spectra) give more readily interpretable spectra as only monomeric species are observed and, for simple molecules such as diatomics, can be analysed totally—including the molar absorptivity of the peaks, just as with UV–vis spectra. However, the great value of infrared spectroscopy to the organic chemist lies in its use as an empirical technique; comparing absorptions of an unknown compound with those amassed by other workers and identifying functional groups by analogy. Nonetheless, to help appreciate the processes involved, it is worth considering the approximate energy needed to cause vibrational excitation of bonds within an organic molecule.

Vibrational frequencies can be calculated with fair precision by using the simple 'ball and spring' model and applying Hooke's Law to correlate frequency with bond strength and atomic masses:

$$v = k\sqrt{\frac{\text{bond strength}}{\text{mass}}} \quad \text{which becomes} \quad v = \frac{1}{2\pi}\sqrt{\frac{k}{\left[\frac{m_1 m_2}{(m_1 + m_2)}\right]}}$$

$v$ = frequency of vibration
$k$ = **force constant** of the bond (the resistance of the bond to vibration and a reflection of the strength of the bond).
$m_1, m_2$ = masses of the two constituent atoms.
$[m_1 m_2/(m_1 + m_2)]$ is the **reduced mass** of the system ($\mu$).

For example, using this relationship, the frequency of the C–H stretching vibration can be calculated from the following data:

$k = 500$ N m$^{-1}$ = $5.0 \times 10^2$ Kg s$^{-2}$ (1 N = 1 kg m s$^{-2}$);
$m_{\text{carbon}} = 20 \times 10^{-27}$ Kg; $m_{\text{hydrogen}} = 1.6 \times 10^{-27}$ Kg

$$v = \frac{1}{2\pi}\left(\frac{5.0 \times 10^2 \text{ Kgs}^{-2}}{(20 \times 10^{-27} \text{ Kg})(1.6 \times 10^{-27} \text{ Kg})/\left[(20 + 1.6) \times 10^{-27} \text{ Kg}\right]}\right)^{1/2}$$

$$v = 9.3 \times 10^{13} \text{ s}^{-1} \text{ (Hz)}$$

However, remembering that organic chemists use infrared spectroscopic analysis in a very empirical manner, it is not necessary to carry out tedious calculations. Instead, working from the above relationship, we should remember the qualitative guideline that vibrational frequency of a bond should increase when bond strength increases and when the reduced mass of the system decreases. From this, the following conclusions may be drawn:

1. Bond stretching requires more energy than bond bending and therefore bond stretching absorptions will require shorter wavelength radiation (higher frequency) than bond bending absorptions.
2. Doubly bonded and triply bonded systems require progressively higher energies for vibrational excitation. Therefore, in either stretching and bending modes, C≡C absorbs higher frequency radiation than C=C which, in turn, absorbs higher frequency radiation than C–C.
3. The smaller the reduced mass of a system, the greater the energy required to cause vibrational excitation and the higher the frequency of radiation necessary. Therefore O–H and C–H stretching occurs at higher frequency than C–O > C–C. Similarly O–H stretching should occur at higher frequency than O–D stretching.

However, this is something of an oversimplification as relative electronegativity of the two constituent atoms (and hence polarization of the bond) is also influential in deciding the overall frequency of absorption. For example, C–H stretching should occur at higher frequency than O–H stretching on the basis of reduced mass, but in fact this is the inverse, as the greater polarization of the O–H bond compared with the C–H bond results in a larger force constant for the bond.

## 3.1  Types of vibrational excitation

Molecules with non-linear arrays of $n$ constituent atoms possess $(3n - 6)$ fundamental vibrational modes. Thus methane possesses 9 different vibrational modes, ethane 18 and so on. However, some of these vibrational modes may be *degenerate*, that is having the same energy, and therefore absorb the same frequency of radiation. In addition, infrared absorptions are only observed if the vibration sets up a fluctuating dipole and therefore symmetrical vibrational modes are not observed in IR spectra; although theoretically symmetrical deformations may still be observed as weak absorptions due to molecular deformation.

In addition to these *fundamental frequencies*, absorptions modulated by the rest of the molecule are commonly observed. *Overtone bands* may be observed at multiples of fundamental vibrations (particularly strong ones) and two fundamental frequencies may combine in either an additive or subtractive way to give *beats* which occur at *combination* or *difference* frequencies.

Evidently, even moderately sized molecules possess a large number of potential vibrational modes and this contributes to the great complexity of IR spectra. For the $XY_2$ unit (e.g. $CH_2$, $NH_2$), the stretching modes may be either *symmetric* or *antisymmetric* and the bending modes are referred to as *scissoring*, *rocking*, *twisting* and *wagging*. Clearly symmetrical vibration modes such as the symmetric stretch and scissoring will only give rise to weak IR absorptions, if at all.

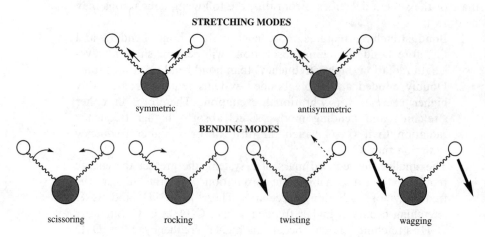

**Fig. 3.2**    Vibrational modes of $XY_2$

### 3.2    Units of measurement

Absorption positions in infrared spectroscopy used to be recorded as the wavelength ν μm and the wavelength scale is still usually shown on all IR spectrum charts; the range of interest to organic chemists being 2.5–25 μm (the *mid-infrared region*). However, the common convention nowadays is to quote absorption positions as *wavenumbers*. The wavenumber is equal to the number of wavelengths per cm (the unit is referred to as the *reciprocal centimetre, cm⁻¹*) and the above frequency range corresponds to 4000–400 cm⁻¹. Commonly chemists erroneously talk of 'frequencies' of $X$ cm⁻¹ rather than correctly talking of wavenumbers of $X$ cm⁻¹ and you should not be confused by this slack nomenclature.

The stretching frequency of the C–H bond is   $\approx 9.3 \times 10^{13}$ Hz
$= 3.3 \ \mu m$
$= 3000 \ cm^{-1}$

## 3.3 The instrument and recording spectra

The principle of operation of the basic double beam IR spectrometer is the same as that of a UV spectrometer; although the higher resolution possible with IR spectra requires more refined optics (Fig. 3.3). The radiation source is an electrically heated filament. This is usually either a Nernst filament (mixed Zr, Th and Ce oxides) or a Globar (silicon carbide). Using mirror optics the beam is split into two. One beam passes through the sample, and the other through the reference—although for certain types of sample no reference is used. After passing through the sample and reference the beams are alternatively passed through a *monochromator* by means of a rotating half silvered mirror and the difference in intensities is recorded by the detector which is simply a thermocouple.

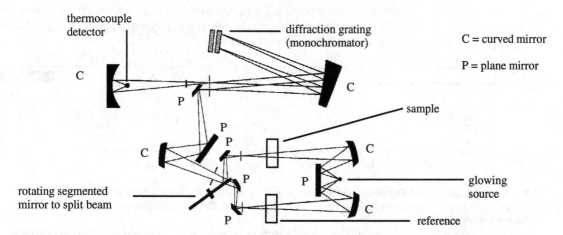

**Fig. 3.3** Optical arrangement of a double beam IR spectrometer (Adapted and reproduced with permission of Perkin Elmer).

The spectrum is recorded over the range 4000–400 cm$^{-1}$ (2.5–25 μm) and is usually divided into two linear sections of 4000–2000 cm$^{-1}$ and 2000–400 cm$^{-1}$, the latter range being expanded compared to the former. The vertical axis is calibrated as *percentage transmittance*, 100% corresponding to no absorption and 0% corresponding to total absorption.

Absorptions are recorded as dips and not peaks in IR spectra.

More recently Fourier transform IR spectrometers have become commonplace in research laboratories. IR radiation, in the form of a complex frequency *interferogram* is passed through the sample, detected and then submitted to Fourier transform. Missing frequencies, due to radiation which has been absorbed, are thus recorded as dips.

## 3.4 Sample preparation

Glass, quartz, and plastic are all opaque to IR radiation and the cells used for containing the sample possess windows made of plates of optically clear sodium chloride. Gas phase spectra are rarely obtained by organic chemists as most samples are liquid or solid. Special cells of 10 cm path length are used to obtain these spectra.

Liquid samples may be readily analysed as a thin film by placing a drop of the neat liquid between two sodium chloride plates and *carefully* squeezing the plates together. The sodium chloride plates are extremely delicate and are easily scratched. Needless to say, they must never be brought into contact with any moisture. Solid samples may also be analysed in this manner as a ***mull*** in a dispersing agent which is commonly a hydrocarbon mixture known commercially as Nujol®. If it is required to observe C–H frequencies in a sample prepared as a mull, hexachlorobutadiene can be used instead. The mull is prepared by grinding the sample in the presence of a few drops of the dispersing agent to produce a solid suspension. The grinding reduces the size of the particles to less than the wavelength of the IR radiation—otherwise the beam would be mainly scattered giving poor resolution. It is usual to run liquid film and Nujol® mulls without a reference sample. Whilst this is relatively unimportant for pure liquid film samples, the mulling agent will impose its absorptions on the spectrum of the material and may obscure absorptions of interest (Fig. 3.4a).

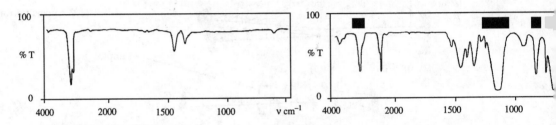

**Fig. 3.4**   (a) IR spectrum of the mulling agent Nujol showing characteristic absorptions at 2919, 2861, 1458 and 1378 cm⁻¹. (b) Spectrum of the commonly used IR solvent chloroform. Areas marked ■ correspond to regions of spectrometer insensitivit running solution spectra.

Solid and liquid samples may both be analysed as solutions (usually in chloroform) in cells specially designed for the purpose (Fig. 3.5). In such cases a cell containing only solvent is placed in the reference beam. The distances between the retaining sodium chloride plates in the solution cells are accurately fixed so the beams pass through the same thickness of solvent (0.1, 0.2, and 1.0 mm are common thicknesses). The solvent absorptions can therefore be subtracted from each other and so only the spectrum of the dissolved sample is recorded. However, chloroform has very strong absorptions at 3020, 1216 and below 759 cm⁻¹ (Fig. 3.4b) making the spectrometer insensitive in these areas due to the inability of the spectrometer to distinguish a small difference between two high absorption intensities. This should be born in mind when analysing aromatic materials in which important diagnostic absorptions occur in the region around 800 cm⁻¹.

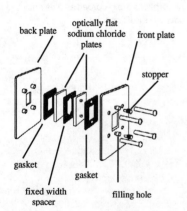

**Fig. 3.5**   The construction of an infrared solution cell. (Adapted and reproduced with permission of Perkin Elmer).

The final method for analysing solid samples is as a potassium bromide disk. In this technique the finely ground sample is mixed intimately with finely powdered KBr and the mixture is squeezed in a press to about 1000 atmospheres. Under these conditions the KBr becomes glassy and forms a thin translucent disc in which the finely ground sample is suspended. This is just like a Nujol mull except for the fact that there are no interfering absorptions due to the mulling agent;

although there may absorptions due to water contained within the KBr which is mildly hygroscopic. The disadvantage of this technique is that it requires a special press and the discs are extremely fragile and can be difficult to mount.

One word of warning: the method of preparation of the sample can have a noticeable effect upon the position of peak absorptions—generally groups capable of intermolecular hydrogen bonding. For example, the carbonyl stretching frequency of ketones and amides is lowered by about (20 cm$^{-1}$ and 40 cm$^{-1}$ respectively) when recorded on the solid or liquid compared with solution samples.

## 3.5  Analysing spectra

### General observations

Broadly speaking the spectrum can be split into three regions, 4000–2000, 2000–1500, 1500–600 cm$^{-1}$ (the later is often referred to as the *fingerprint region*). Most interpretable data are contained in the first two regions. The fingerprint region contains absorptions due to the molecular vibrations of the compound and rigorous assignment in this region is not possible. The 4000–1500 cm$^{-1}$ region contains bond stretching absorptions. In all analyses the pattern to follow is essentially the same:

1. Inspect the spectrum starting at high wavenumbers (4000 cm$^{-1}$).
2. Note which are the strongest absorptions and attempt to correlate them from tables.
3. Note absence of peaks in important areas.
4. Do not attempt to correlate all peaks, especially in the fingerprint region.

It takes time to learn even a moderate range of correlation data but you should make an effort to memorize the brief table in the Appendix which contains the most important functional group absorptions broken down according to wavenumber of absorption. Note that all absorption positions quoted are approximate.

### Detailed analysis

#### 4000–2500 region

This is the region where bonds to hydrogen usually absorb (due to low reduced mass of X–H). Beware of overtone peaks of strong carbonyl absorptions around 1750 cm$^{-1}$.

Hydroxyl peaks are usually observed as broad Gaussian-shaped peaks centred around 3500 cm$^{-1}$ due to intermolecularly hydrogen-bonded hydroxyl groups. Absorptions due to monomeric hydroxyls are less commonly seen, appearing as sharp absorptions at about 3600 cm$^{-1}$. *Intra*molecularly hydrogen-bonded absorptions can be distinguished from *inter*molecularly hydrogen-bonded absorptions by progressively diluting the solution (Fig. 3.6). Notice how the broad stretching absorption due to the intermolecular H–bonded OH is progressively replaced by a sharp peak due to monomeric species (although, of course, there is also an overall decrease in intensity as the solution becomes more dilute).

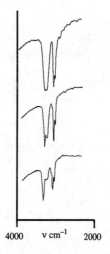

4000    v cm$^{-1}$    2000

**Fig. 3.6**  Progressively diluting a solution of an alcohol results in the broad intermolecular H–bonded OH stretching absorption centred on 3500 cm$^{-1}$ being replaced with a sharper absorption at 3600 cm$^{-1}$ due to the monomeric species. Note also the diminution of intensity with increasing dilution.

Carboxylic acids readily hydrogen-bond and the OH peak of these is particularly broad due to the strong tendency of acids to form hydrogen-bonded dimers (Fig. 3.7 b).

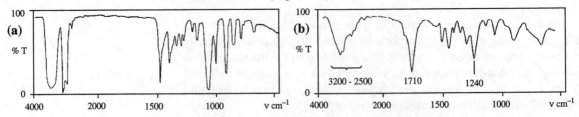

**Fig. 3.7**   Solution spectra of (a) cyclohexanol, (b) propanoic acid showing the different appearance of O–H stretching absorptions.

Amine N–H absorption can occur in the same area as O–H and may cause confusion. However, O–H stretching is always accompanied by a C–O stretch near 1250 cm$^{-1}$ (and C=O stretch at 1730 cm$^{-1}$ for carboxylic acids). N–H stretching absorptions are usually less intense than O–H and usually show two (primary amines, Fig. 3.8) or one (secondary amines) sharp spikes superimposed on the broad peak as well as N–H bending around 1600 cm$^{-1}$.

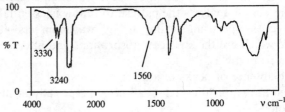

**Fig. 3.8**   Spectrum of hexylamine (film) showing asymmetric and symmetric NH$_2$ stretch.

Primary and secondary amides also possess N–H bonds and show absorptions in this region, but an amide can be confirmed by the presence of a carbonyl absorption around 1690 cm$^{-1}$ (*amide I*). Primary and secondary amides also show a strong N–H bending absorption (*amide II*) in the 1690–1520 cm$^{-1}$ region. The amide I and II positions depend upon whether the sample is analysed in solution, or as a Nujol® mull or KBr disc (Fig. 3.9). The amide I absorption is about 40 cm$^{-1}$ lower and the amide II about 30 cm$^{-1}$ higher in Nujol® and KBr spectra compared to solution spectra.

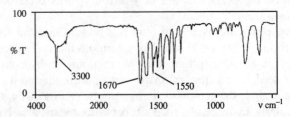

**Fig. 3.9**   Spectrum of *N*–phenyl ethanamide (KBr disc) showing NH stretch and amide I and II.

C–H stretching absorptions occur in the range 3300–2700 cm$^{-1}$ and as almost all organic molecules contain C–H bonds this is generally not particularly diagnostic. However, C–H bonds of terminal alkynes show very sharp strong absorptions at 3300 cm$^{-1}$ and can be distinguished from monomeric OH absorptions, with which they might be confused, as they are accompanied by an absorption around 2100 cm$^{-1}$ due to the C≡C stretch (Fig. 3.10a). The stretching absorptions of C–H bonds attached to sp$^2$ hybridized carbons come at about 3100 cm$^{-1}$ and are usually relatively weak (Fig. 3.10b), particularly those attached to aromatic carbons. In fact, lack of a strong absorption in this region might suggest the presence of an aromatic material.

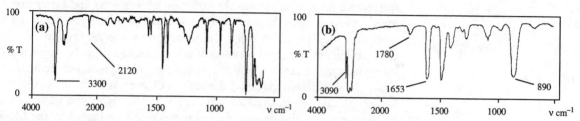

**Fig. 3.10**   Spectra of (a) phenylethyne (b) 2-methyl-1-heptene showing absorptions associated with C≡CH and C=CH.

Aldehyde C–H stretching usually shows two absorptions in the region 2900–2700 cm$^{-1}$, but the presence of an aldehyde must always be confirmed by the presence of a C=O stretch.

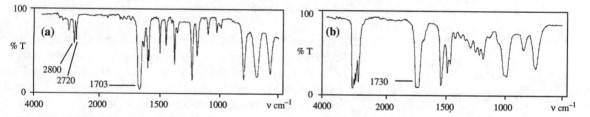

**Fig. 3.11**   Solution spectra of (a) benzaldehyde (b) propanal.

## 2500–1900 region

Basically the only peaks occurring in this region are due to absorptions resulting from triple and cumulated double bond systems. As mentioned already, disubstituted alkynes absorb at 2200–2100 cm$^{-1}$. With dialkylated alkynes, having a dipole moment close to zero, the absorption is usually very weak and may be absent altogether; although substitution with electronically dissimilar substituents increases intensity and lowers the frequency of absorption (Fig. 3.10a). Cyanides show a characteristically strong absorption at 2260–2200 cm$^{-1}$ (Fig. 3.12).

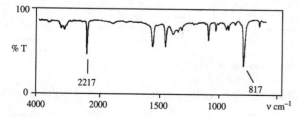

**Fig. 3.12**   Spectrum of cyano-4-methylbenzene showing CN stretch at 2217 cm$^{-1}$ and 1,4-disubstituted benzene C–H bending at 817 cm$^{-1}$.

Don't forget that $CO_2$ absorbs strongly in this region and may give a spurious peak if path imbalances arise, particularly with liquid film samples when a reference is usually not used.

## 1900–1500 region

The important species absorbing in this region are C=O and C=C. Due to the large dipole moment of the carbonyl bond, C=O stretching results in very intense absorptions—usually the most intense in the whole spectrum. Carboxylic acids usually give the strongest absorptions and esters absorb more strongly than ketones, aldehydes and amides, the latter being somewhat variable in intensity. Carbonyl groups absorb around the 1700 cm$^{-1}$ region and the exact position of absorption can often be diagnostic of the actual type of carbonyl group responsible. Some general guide-lines are:

1.   In substituted carbonyl derivatives of general formula RCOX, the higher the electronegativity of substituent X, the higher the wavenumber of the carbonyl absorption. Anhydrides show two bands in the 1850–1740 cm$^{-1}$ region. The higher wavenumber band is stronger in acyclic anhydrides; whereas the inverse is true for cyclic anhydrides. Acid chlorides absorb in the 1815–1790 cm$^{-1}$ region with acid bromides absorbing at higher wavenumber and iodides at lower. Esters absorb around 1750–1735 cm$^{-1}$ ketones around 1725–1705 cm$^{-1}$ and aldehydes around 1720–1700 cm$^{-1}$ (the latter some 20 cm$^{-1}$ higher if in solution, Fig. 3.11b).

2.   If the carbonyl is part of a ring smaller than 6-membered, decreasing the size of the ring leads to the absorption moving to higher wavenumber due to enforced compression of the sp$^2$ angle of the carbonyl carbon. Rings which are 6-membered or larger behave as acyclic compounds. This effect holds for both cyclic ketones and lactones (Fig. 3.13).

3.   α,β-Unsaturation, either a double bond or an aromatic ring, lowers the absorption wavenumber by about 40–45 cm$^{-1}$ (Figs 3.11a, 3.14b). Unlike in UV–visible spectra, additional extension of the conjugation does not affect the position further. The important exceptions to this general rule are amides which are shifted to higher frequency by about 15 cm$^{-1}$. In all of these instances, the carbonyl absorption will be accompanied by a strong double bond stretching absorption at about 1650 cm$^{-1}$.

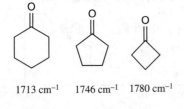

1713 cm$^{-1}$   1746 cm$^{-1}$   1780 cm$^{-1}$

**Fig. 3.13** Decreasing ring size causes an increase in wavenumber for the carbonyl stretching absorption.

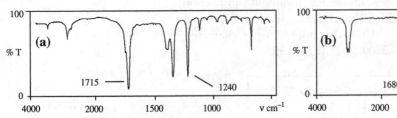
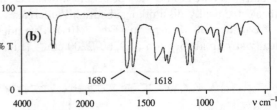

**Fig. 3.14**   Spectra of (a) propanone (film) (b) 2-methylpent-2-enone (mesityl oxide) (film).

4.   Intramolecular hydrogen bonding lowers the wavenumber by 50 cm$^{-1}$.

5.   Inductively electron withdrawing substituents at the α-position cause a shift to higher wavenumber of about 20 cm$^{-1}$ for each substituent.

6. The position of carbonyl absorption in liquid film or solid state spectra of aldehydes and ketones is about 20 cm$^{-1}$ lower compared with solution spectra.

Alkene double bond absorptions are found in the 1680–1500 cm$^{-1}$ region with symmetrical alkenes giving very weak absorptions. Conjugation with a carbonyl group gives stronger absorption at around 1650 cm$^{-1}$ but this is always less intense than the carbonyl absorption. Enamines and enol ethers show increased intensity and absorb at the higher frequency end of the range. Aromatic rings show two or three absorptions in the 1600–1500 cm$^{-1}$ range.

### The fingerprint region

This is a complicated region and it is rarely possible to assign many of the peaks within it. Its name is derived because it may be used for direct identification by comparison of an authentic material with a substance obtained from another source. There are some useful absorptions however. The appearance of a C–O stretch at 1150–1070 cm$^{-1}$ permits distinction between esters and ketones. Nitro- and sulfonyl groups give a pair of strong absorptions between 1550–1350 cm$^{-1}$. Finally, the degree of substitution of aromatic rings can often be determined from the sets of strong bands in the 850–730 cm$^{-1}$ region which result from C–H out-of-plane bending.

| | |
|---|---|
| Monosubstituted | 770–730, |
| | 720–680 cm$^{-1}$ |
| (see Fig. 3.11a) | |
| 1,2-disubstituted | 770–735 cm$^{-1}$ |
| 1,3-disubstituted and 1,2,3-trisubstituted | |
| | 810–750 cm$^{-1}$ |
| 1,4-disubstituted | 860–800 cm$^{-1}$ |
| (see Fig. 3.12) | |

Absorptions relating to benzene substitution patterns.

## 3.6 Exercises

1. The following are spectra of benzyl alcohol and 1-octene recorded in chloroform. Identify each spectrum and assign the major absorptions.

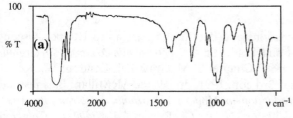
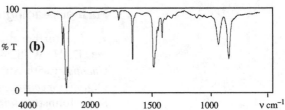

2. The four following spectra were recorded on solutions of an acyl chloride, an aldehyde, a carboxylic acid and an amide. Which corresponds to which?

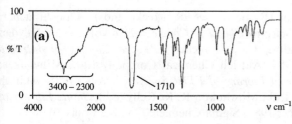
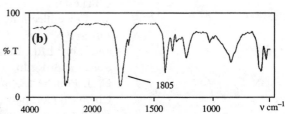

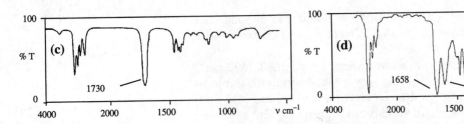

3.   Decide which of the following spectra corresponds to butyrolactone, cyclohexanone, ethyl ethanoate and methyl phenyl ketone, and assign the labelled peaks.

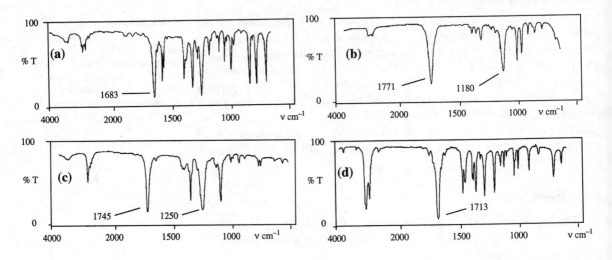

## Further reading

1.   For further discussions and data tables relating to IR spectroscopy see: D. H. Williams and I. Fleming, *Spectroscopic Methods in Organic Chemistry*, 5th edn, 1995, Chapter 2, McGraw-Hill London; W. Kemp, *Organic Spectroscopy*, 3rd edn, 1991, pp. 19–98, McMillan, Basingstoke; S. F. Johnston, *Introduction to Modern Vibrational Spectroscopy*, 1991, Ellis Horwood, Chichester; N. P. G. Roeges, *A Guide to Complete Interpretation of Infrared Spectra of Organic Structures*, 1994, John Wiley and Sons Ltd, Chichester.

2.   For a discussion of the principles and practical aspects of Fourier-transform infrared spectroscopy see: S. F. Johnston, *Fourier-Transform Infrared: a Constantly Evolving Technology*, 1991, Ellis Horwood, Chichester.

3.   For collections of IR spectra see: R. Mecke and F. Langenbucher, *Infrared Spectra of Selected Chemical Compounds*, 8 vols., 1965, Heyden and Son Ltd, London; C. J. Pouchert, *Aldrich Library of Infrared Spectra*, 3rd edn, 1984, Aldrich Chemical Company Ltd., Milwaukee; C. J. Pouchert, *Aldrich Library of FTIR Spectra*, 3 vols., 1985, Aldrich Chemical Company Ltd., Milwaukee, R. K. Kelly (ed.), *Sigma Library of FTIR Spectra*, 2 vols., 1986, Sigma Chemical Co., St. Louis.

# 4 Nuclear magnetic resonance spectroscopy: the basics

There can be no doubt that nuclear magnetic resonance (NMR) spectroscopy is now the most powerful and versatile of all analytical techniques used routinely by organic chemists. It is now fifty years since the phenomenon was first observed experimentally, and, within only five years of this, it was being used to address chemical problems. The usefulness of this technique in chemistry can be attributed largely to the very detailed information that can be obtained; spectroscopic features correlate with individual atoms within a molecule, rather than groups of atoms as in UV or IR spectroscopy. Furthermore, these features relate, in a fairly straightforward and direct manner, to the chemist's representation of chemical structure and bonding, as we shall discover in this chapter, and it is now routinely possible to determine the structures of small- to- medium-sized organic molecules (with a molecular mass of less than approximately 1000 daltons) in solution. Although NMR spectroscopy of solids is a well established field, we shall consider only solution spectroscopy as it is in this form that NMR finds greatest use in organic chemistry.

## 4.1 The physical background to NMR

The phenomenon of nuclear magnetic resonance occurs because the nuclei of certain atoms possess *spin*. The spin is characterized by the nuclear spin quantum number, $I$, which may take integer and half-integer values ($I = \frac{1}{2}$, 1, $\frac{3}{2}$, 2 and so on). From the chemist's point of view, it is generally the nuclei that have $I = \frac{1}{2}$ that are of most interest because those with $I$ greater than $\frac{1}{2}$ turn out to have unfavourable properties with regard to the observation of their spectra. Nuclei with zero spin ($I = 0$) are not amenable to NMR observation, although most elements have at least one isotope that does possess nuclear spin. In organic chemistry, the most commonly studied nucleus is that of $^1$H, followed by $^{13}$C. You may find it surprising that the carbon nucleus is *not* the most popular one for observation, and this is because the most abundant carbon isotope, $^{12}$C, has zero spin (as do all nuclei with atomic and mass numbers both even), so we are generally forced to observe the low natural abundance (1.1%) of $^{13}$C and face problems with one of the major limitations in NMR, that of sensitivity. The $^1$H nucleus, however, has almost 100% natural abundance and is one of the most sensitive nuclei to observe by NMR. Other commonly encountered nuclei are given in Table 4.1, together with some of their relevant parameters, which are discussed below.

### Nuclei in magnetic fields

Some nuclei may possess spin and therefore have angular momentum, and

**Table 4.1**  Properties of NMR active isotopes of commonly encountered nuclides. Resonance frequencies are for a field strength of 11.74 tesla and the relative sensitivity includes a term for the intrinsic sensitivity of a nuclide and its natural abundance.

| Nucleus | Natural Abundance % | Nuclear Spin (I) | Magnetogyric ratio $/10^7$ rad $T^{-1}s^{-1}$ | Resonant frequency / MHz | Relative sensitivity |
|---|---|---|---|---|---|
| $^1H$ | 99.98 | $\frac{1}{2}$ | 26.75 | 500.00 | 1.00 |
| $^2H$ | 0.02 | 1 | 4.11 | 76.75 | $1.45 \times 10^{-6}$ |
| $^{13}C$ | 1.11 | $\frac{1}{2}$ | 6.73 | 125.72 | $1.76 \times 10^{-4}$ |
| $^{14}N$ | 99.63 | 1 | 1.93 | 36.12 | $1.01 \times 10^{-3}$ |
| $^{15}N$ | 0.37 | $\frac{1}{2}$ | −2.71 | 50.66 | $3.85 \times 10^{-6}$ |
| $^{17}O$ | 0.04 | $\frac{5}{2}$ | −3.63 | 67.78 | $1.08 \times 10^{-5}$ |
| $^{19}F$ | 100.00 | $\frac{1}{2}$ | 25.18 | 470.39 | 0.83 |
| $^{31}P$ | 100.00 | $\frac{1}{2}$ | 10.84 | 202.40 | $6.63 \times 10^{-2}$ |

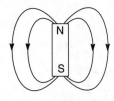

**Fig. 4.1**  Spin-$\frac{1}{2}$ nuclei may be considered to act as microscopic bar magnets.

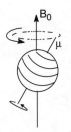

**Fig. 4.2**  Spinning nuclei precess in an applied magnetic field, $B_0$ at their characteristic Larmor frequencies.

Spectrometers are usually referred to by the frequency at which protons would resonate if placed in the instrument, for example, we may speak of a '200 MHz spectrometer' or a '500 MHz spectrometer'.

all nuclei have charge. The motion of this charge, that is, the spinning, means that these nuclei have associated with them a weak magnetic field (they possess a **magnetic moment**, μ); indeed the nuclei can be considered to act as small bar magnets (Fig. 4.1). When these nuclei are placed in an external magnetic field (designated $B_0$), they experience a torque which forces them into precession about the axis of the external field (Fig. 4.2). This motion is referred to as **Larmor precession** and it occurs at the **Larmor frequency**, ν, which is directly proportional to the strength of the applied magnetic field. The precession is analogous to the motion of a spinning gyroscope in the Earth's gravitational field; the central wheel of the gyroscope spins about its own axis whilst this axis, in turn, precesses about the gravitational field. The Larmor frequency also depends on the **magnetogyric ratio**, γ, of the nucleus (still often referred to as the **gyromagnetic ratio**, although this is not S.I. terminology). This is defined as

$$\gamma = \frac{magnetic\ moment}{angular\ momentum}\ \text{rad T}^{-1}\ \text{s}^{-1}$$

This value may be considered to be an indication of how 'strongly magnetic' a nucleus is and its value is constant for a given nuclide ($^1H$, $^2H$, $^{13}C$ etc., see Table 4.1). The Larmor frequency is then given by

$$\nu = \frac{\gamma B_0}{2\pi}\ \text{Hz}$$

Thus, the nuclei possess a magnetic moment which is acted upon by the external field. Nuclei can take up $2I + 1$ possible orientations in this field so those with $I = \frac{1}{2}$ may align themselves in two ways; parallel to the field or anti-parallel to it, the former being of (slightly) lower energy. There are only these discrete orientations possible because the energy levels involved are **quantized**, with the lower energy state generally being designated α, the higher β. In common with UV and IR spectroscopy, the lower energy α state may be excited to the higher level by the application of electromagnetic radiation oscillating at the appropriate frequency and therefore induce **nuclear magnetic resonance** (Fig. 4.3). Likewise, the higher energy β state may lose its excess energy by a suitable means (known as relaxation, see

Section 5.2) and fall to the α state; clearly in both cases this simply corresponds to a nuclear spin inverting its orientation with respect to the field. The frequencies required to excite these transitions correspond to the precessional frequencies of the nuclei, that is, their Larmor frequencies. With currently available magnetic fields (1.4 to 17.5 tesla) these fall in the range 60–750 MHz, which is within the radiofrequency region of the spectrum, the energies involved being given by

$$\Delta E = h\nu = \frac{h\gamma B_0}{2\pi}$$

$h$ = Planck's constant
$\nu$ = Larmor frequency
$B_0$ = applied field strength
$\gamma$ = magnetogyric ratio

The need for relatively low frequency radiation indicates the energies between the α and β states to be very small. When we are recording a spectrum, we are actually sampling the population differences between these energy states, the differences at equilibrium being defined by the Boltzmann equation

$$\frac{N_\alpha}{N_\beta} = e^{\Delta E/RT}$$

$N$ = number of nuclei in spin state
$\Delta E$ = energy of transition
$R$ = gas constant
$T$ = absolute temperature

In NMR spectroscopy, these population differences are only in the order of 1 part in $10^4$ even with the strongest available field strengths, which means NMR is a relatively insensitive technique and sample size requirements are high; it is not uncommon for sample quantities to be within the milligram region, some orders of magnitude greater than those in optical spectroscopy.

Within a molecule the energy levels associated with different nuclear environments will vary, as we shall see shortly, and so produce spectra displaying various resonance frequencies. To obtain a spectrum containing all these frequencies within a given sample one might suppose that a sensible approach would be to follow the procedures adopted in optical spectroscopy. Thus, one would place the sample in a magnetic field, irradiate it with a radiofrequency and scan through either a radio-frequency range or a magnetic field range, recording the frequencies at which resonance occurred. This was indeed how earlier spectrometers (built prior to approximately 1980) would operate and the scheme is known as *continuous wave* NMR spectroscopy operating in either frequency-sweep or field-sweep mode. However, this mode of operation is inferior to the modern pulse-Fourier transform (FT) approach described in Chapters 1 and 5 and, in time, will become redundant. We shall consider FT NMR methodology only.

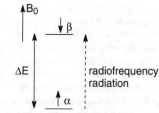

**Fig. 4.3** Spin–$\frac{1}{2}$ nuclei may take up two orientations with respect to the applied field; parallel (α) or antiparallel (β) to it. **Nuclear magnetic resonance** occurs when radiation of the appropriate frequency is applied.

## Instrumentation

A schematic representation of a modern FT NMR spectrometer is presented in Fig. 4.4. The sample solution is held in a cylindrical glass tube and is placed in the magnet. The tube sits within a coil of wire (essentially a radiofrequency antenna) which is tuned to the nuclear frequency of interest, in much the same way that one would tune a radio receiver to the desired radio station. The pulse of radiofrequency radiation, typically a few µs in duration, from a suitable transmitter is delivered *via* this coil, which also receives the emitted frequencies from the stimulated sample in the form of an oscillating voltage (see Section 5.1). A reference frequency, that of the original pulse, is subtracted to reveal *differences* between the stimulating and detected frequencies, which are up to a few tens of

The detected frequencies, after subtraction of a **reference** or **carrier frequency**, are in the audio range, so it is also possible to *hear* NMR spectra with suitable hardware,- a loudspeaker!

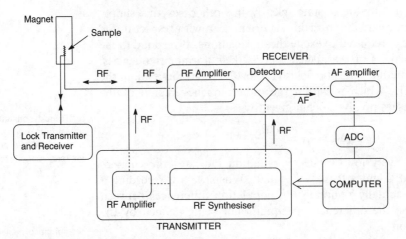

**Fig. 4.4**   Schematic representation of a Fourier transform NMR spectrometer. Abbreviations are RF, radiofrequency, AF, audiofrequency, and ADC, analogue-to-digital converter.

kilohertz; in the audiofrequency range. These signals may then be converted to digital values *via* an analogue-to-digital converter (ADC) and then stored in the computer for signal averaging and analysis.

It is not uncommon to wish to resolve frequency differences of less than 1 Hz, requiring the magnetic field to be uniform and stable to 1 part in $10^9$. This remarkable feat is achieved by the use of superconducting solenoid magnets, which are very stable, and by spinning the sample whilst in the magnet, which averages away small field inhomogeneities that may remain. The ***field-frequency lock*** system monitors the deuterium resonance of the solvent and corrects any drift in magnetic field that may occur. This is one of two reasons for using ***deuterated*** solvents for NMR spectroscopy. The second is the simple fact that because there is usually far more solvent than solute in the NMR tube, the signals of interest would be swamped by those of a *protonated* solvent, whereas deuterium signals will not be apparent in a proton spectrum. Deuterated chloroform is the most widely used solvent, and the properties of this and other common solvents are presented in Table 4.2.

### Features of the $^1$H spectrum

The $^1$H NMR spectrum of ethyl *p*-tolylacetate is shown in Fig. 4.5. It comprises essentially three features:

1.  Proton resonances are distributed along the frequency axis. Each proton sits in a distinct chemical environment which is character-ized by its ***chemical shift***, $\delta$ (the units of which will be described below).
2.  Different peaks in the spectrum can be seen to be present with differing intensities, which relate to the numbers of protons giving rise to the signal.
3.  Some resonances can be seen to possess fine structure, due to the proton interacting with neighbouring atoms. The degree of interaction, or coupling, is characterised by a ***coupling constant***, *J*, which has units of hertz (Hz).

These three features characterize *all* one-dimensional NMR spectra and

High quality sample preparation is essential for high quality NMR spectra. This should not come as a surprise when one considers the need to maintain a magnetic field throughout the sample that is homogeneous to 1 part in $10^9$! Thus, samples must contain no solid particles ***at all*** as these will distort local magnetic fields, sample tubes must not be scratched or cracked and must be kept clean and dry. Do not use chromic acid for cleaning tubes as residual traces will have detrimental effects on spectra (broad lines result from paramagnetic chromium, see Section 5.2). Maximize sample concentration by using the minimum solvent volume that is consistent with your spectrometer configuration.

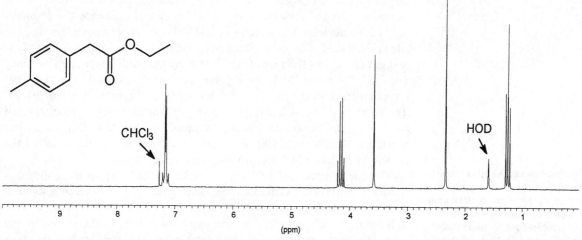

**Fig. 4.5** The $^1$H spectrum of ethyl *p*-tolylacetate.

an understanding of these and how they are influenced provides the basis for the interpretation of any spectrum, and each is considered in the following sections. The spectrum is referred to as being one-dimensional as it possesses one *frequency* axis. More advanced techniques may utilise two frequency axes and are therefore referred to as two-dimensional experiments which are introduced in the following chapter (Section 5.5).

## 4.2 Chemical shifts

When a nucleus is placed in an external magnetic field, it precesses at a rate determined by the magnetogyric ratio of the nucleus and the strength of the magnetic field. The nucleus is also surrounded by an electron cloud which will also be circulating in the magnetic field, about the nucleus. The

**Table 4.2** Properties of the common deuterated solvents. Proton shifts, $\delta_H$, and carbon shifts $\delta_C$, are quoted relative to tetramethylsilane, TMS (proton shifts are those of the residual partially ***protonated*** solvent). The proton shifts of residual HOD ($\delta_{(HOD)}$) vary depending on solution conditions.

| Solvent | $\delta_H$ /ppm | $\delta_{(HOD)}$ /ppm | $\delta_C$ /ppm | Melting Point /°C | Boiling Point /°C |
|---|---|---|---|---|---|
| Acetone-d$_6$ | 2.05 | 2.0 | 206.7, 29.9 | −94 | 57 |
| Acetonitrile-d$_3$ | 1.94 | 2.1 | 118.7, 1.4 | −45 | 82 |
| Benzene-d$_6$ | 7.16 | 0.4 | 128.4 | 5 | 80 |
| Chloroform-d$_1$ | 7.27 | 1.5 | 77.2 | −64 | 62 |
| Deuterium oxide-d$_2$ | 4.80 | 4.8 | − | 3.8 | 101 |
| Dichloromethane-d$_2$ | 5.32 | 1.5 | 54.0 | −95 | 40 |
| N,N-dimethyl formamide-d$_7$ | 8.03, 2.92, 2.75 | 3.5 | 163.2, 34.9, 29.8 | −61 | 153 |
| Dimethylsulfoxide-d$_6$ | 2.50 | 3.3 | 39.5 | 18 | 189 |
| Methanol-d$_4$ | 4.87, 3.31 | 4.9 | 49.2 | −98 | 65 |
| Tetrahydrofuran-d$_8$ | 3.58, 1.73 | 2.4 | 67.6, 25.4 | −109 | 66 |
| Toluene-d$_8$ | 7.09, 7.00, 6.98, 2.09 | 0.4 | 137.9, 129.2, 128.3, 125.5, 20.4 | −95 | 111 |

circulating electrons generate magnetic fields of their own, which act in opposition to the applied external field and will influence the resultant field at the nucleus. The nucleus is said to be **shielded** to some extent from the applied field. The small change experienced at the nucleus will produce a slight change in the rate at which it precesses, and hence the frequency required to stimulate it into resonance. Differences in the chemical environment modify the electron density and distribution about nuclei. The corresponding different degrees of shielding can therefore be expected to produce resonances at differing frequencies. These differences are what constitute the chemical shift differences we observe and this is one of the means by which NMR spectra may be related to chemical structure.

The effects produced by the local, internal fields, are considerably less than that imposed by the external field. The latter produces resonance frequencies at many hundreds of megahertz, whereas the former produce differences which are only of the order of a few kilohertz. From the chemist's point of view we are interested in the frequency differences between nuclei in differing chemical environments, and so it turns out to be convenient to define the frequency scale as a **relative** scale. The **chemical shift**, $\delta$, for any resonance is defined as:

$$\delta = \frac{\nu - \nu_{ref}}{\nu_{ref}}$$

Strictly, this is a dimensionless property, but as the result is of the order of $10^{-6}$, the units of parts per million (ppm) are used so as to make this value implicit. The reference compound for both $^1$H and $^{13}$C NMR is tetramethylsilane (TMS) whose chemical shift is, therefore, defined as 0.0 ppm. This is a readily available, inert, and volatile liquid which is soluble in the majority of solvents and displays a single resonance whose shift falls conveniently in a region away from the vast majority of commonly encountered resonances. Aside from making the numbers involved convenient to handle, the use of the relative scale has a further benefit; **the chemical shift of any resonance, in ppm, is independent of the strength of the applied magnetic field** despite the fact that the frequency differences in Hz between the resonances will alter in direct proportion to the field. This means that any chemical shift assignments made on one spectrometer are equally applicable to the spectra obtained at any field strength. Fig. 4.6 displays $^1$H spectra obtained at a variety of field strengths. The chemical shifts for the resonances do not alter with increasing field, although their actual frequency differences do increase, that is, **greater peak dispersion is obtained at higher field strengths.**

Resonances at higher frequency are often referred to as being **deshielded** or shifted to low-field whereas those at lower frequencies are described as **shielded** or shifted to high-field (Fig. 4.7). The references to high- and low-field arise from terminology used in the continuous wave era although they are still in widespread use.

Now that we have seen what constitutes the chemical shift, we shall go on to see what influences this property and how this relates to chemical structures. Primarily, **intramolecular effects** result in substantial changes in chemical shifts, so, in most cases, the differences in shifts that are observed can be related to the structural features within a single molecule. Figure 4.8 displays the shift regions in which various proton resonances are

The $\delta$ scale has not always been used for defining chemical shifts, and in older literature you may encounter the $\tau$ scale, although this should never be seen in modern publications. For proton NMR the relationship is $\delta = 10 - \tau$, so values may be readily altered.

$\nu$ = Frequency of the resonance
$\nu_{ref}$ = Frequency of the resonance of a reference compound

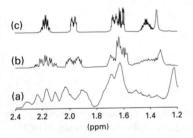

**Fig. 4.6**  Partial $^1$H spectra recorded at frequencies of (a) 90 (b) 200 and (c) 500 MHz.

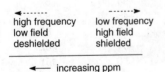

**Fig. 4.7**  Nomenclature used when describing relative chemical shifts.

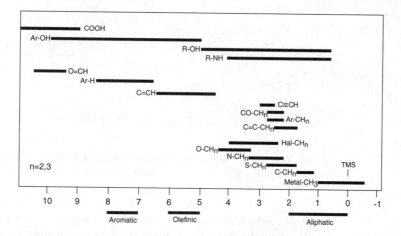

**Fig. 4.8** Typical proton chemical shift ranges for various chemical environments. The ranges below the table show where alkane, alkene and arene protons would be expected to resonate in the absence of significant substituents.

commonly encountered, and more detailed tables for predicting chemical shifts may be found in the Appendices. One generality to be aware of is that increased branching leads to an increase in chemical shift, so that for any functional group we find $\delta$ CH $>$ $\delta$ CH$_2$ $>$ $\delta$ CH$_3$.

## Inductive effects

The degree of shielding about any nucleus will be strongly influenced by the electron density surrounding it, which, in turn, will be affected by the presence of electronegative and electropositive neighbours. Electron withdrawing groups will tend to draw electron density away from the nucleus, so reducing the shielding effect and moving the resonance frequency to higher values (greater chemical shift, Fig. 4.9); the opposite arguments apply to electropositive groups.

The effect tends to fall rapidly as the nucleus becomes more distant from the inductive group. For example, in proton spectroscopy, effects may be readily observed over two and, to a lesser extent, three bonds, but are often vanishingly small beyond this (Fig. 4.9).

The effects of multiple substituents tend to be approximately additive, and tables may be used which allow one to predict the shifts of a nucleus in a given environment (see Appendices). The advent of more sophisticated NMR techniques means that chemists have to rely less on empirical estimates of chemical shifts to be able to assign spectra, and the need to pore over such tables is corresponding less. An understanding of general trends, such as those shown in Fig. 4.8 is, however, fundamental to the routine analysis of NMR spectra and is strongly encouraged.

## Anisotropic effects

Often, the distribution of electron density within chemical bonds is unsymmetrical, and the shift of any nucleus in the vicinity of the bond will be dependent on its position relative to it. This *anisotropic effect* is most pronounced in unsaturated systems, in which the $\pi$-electrons circulate locally in response to the external field and give rise to shielding effects that are highly spatially dependent. Whether this effect causes shifts to

$XC^1H_2C^2H_2C^3H_3$

| X | $C^1H_2$ | $C^2H_2$ | $C^3H_3$ |
|---|---|---|---|
| Et | 1.3 | 1.3 | 0.9 |
| HOOC | 2.3 | 1.7 | 1.0 |
| SH | 2.5 | 1.6 | 1.0 |
| NH$_2$ | 2.6 | 1.5 | 0.9 |
| Ph | 2.6 | 1.6 | 0.9 |
| Br | 3.4 | 1.9 | 1.0 |
| Cl | 3.5 | 1.8 | 1.0 |
| OH | 3.6 | 1.6 | 0.9 |
| NO$_2$ | 4.4 | 2.1 | 1.0 |

**Fig. 4.9** The influence of electronegative substituents on chemical shift (in ppm). The greater the electron withdrawing effect of the substituent, the greater the chemical shift of adjacent protons.

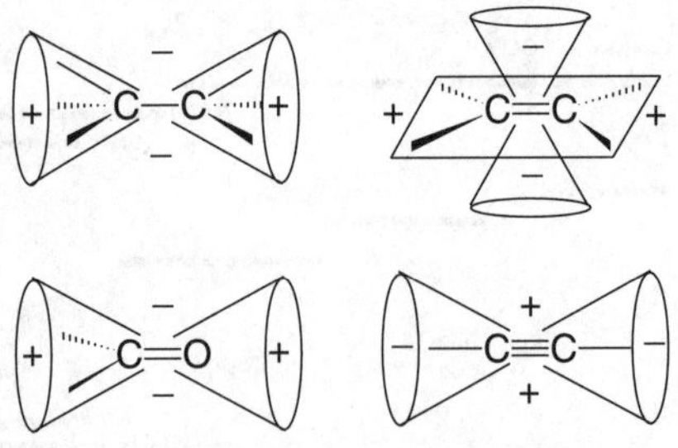

1.0 1.7 2.4
$CH_3CH_2CH_2$—C=O
|
H
9.8

0.9 1.4 2.0
$CH_3CH_2CH_2$    H
\    /
C=C
/    \
H    $CH_2CH_2CH_3$
5.4

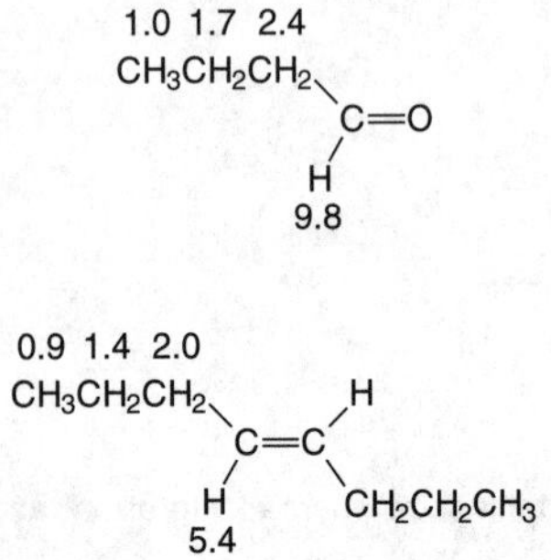

**Fig. 4.11** Anisotropy is a localized effect. Only protons directly attached to the π system experience significance deshielding.

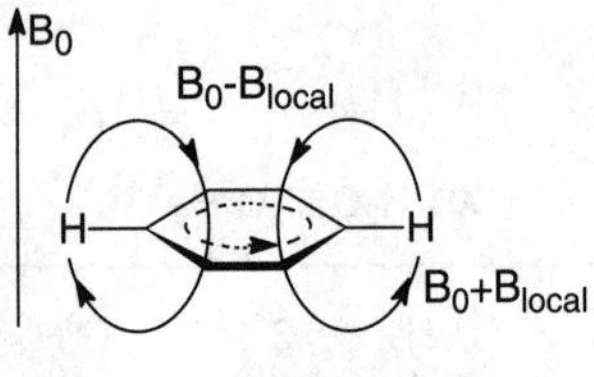

**Fig. 4.12** The ring current effect in benzene moves aromatic proton resonances to higher frequency.

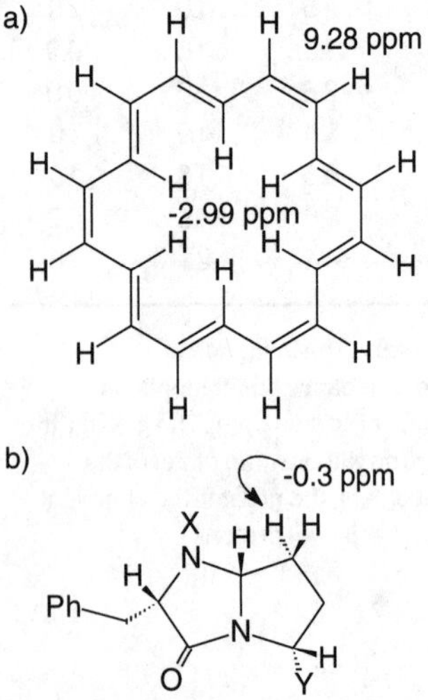

a)

b)

**Fig. 4.13** (a) Ring current effects in 18-annulene cause dramatic shift differences for the 'inner' and 'outer' protons. (b) Unusual shifts can be useful in predicting stereochemistry.

**Fig. 4.10** Chemical shift anisotropy caused by some single, double and triple bonds. (+) indicates deshielding regions (increased chemical shift) whereas (−) indicates shielding regions (decreased chemical shift).

higher or lower frequency will depend not only on position, but also on the atoms in the bond(s) in question. The trends that are observed are most conveniently represented by regions of shielding or deshielding about the bond (Fig. 4.10).

This effect explains why resonances of alkene protons fall typically in the 5–6 ppm region, whereas those of alkyne protons are found between 2–3 ppm. The anisotropy of the C=O bond is somewhat larger and aldehyde resonances are found, characteristically, around 9–10 ppm. In any case, the effect falls rapidly with distance from the π-system (Fig. 4.11) so the observation of protons with high-frequency shifts often suggests they are part of these functional groups.

The hybridization of the carbon atom also has some influence on the shift of the attached proton as the electronegativity of the carbon increases as it takes on more s-character.

An additional influence is found in the case of aromatic rings. The proton resonances of benzene, for example, occur at *ca.* 7.3 ppm, so are considerably less shielded than protons in a lone double bond. This is attributed to the **ring-current** caused by the circulation of the π-electrons around the ring on the application of an external field. The circulating electrons produce a field ($B_{Local}$) that adds to the external field at the position of the protons and so causes shifts to higher frequency (Fig. 4.12). Conversely, the regions above and below the plane of the ring experience shielding from the applied field, and nuclei in these regions can be expected to resonate at lower frequencies. Such effects are indeed observed in the spectra of annulenes ($(4n + 2)$ π systems, Fig. 4.13a) for which, in the case of 18-annulene, the 'inner' and 'outer' proton shifts are −2.99 and 9.28 ppm respectively. The effect also has some use in stereochemical analysis, such as for the example of Fig. 4.13b, in which the unusual shift of −0.3 ppm, suggests the proton sits directly above the plane of the phenyl ring, which has been shown to be the case.

Proton resonances in the 7–8 ppm region are often indicative of an aromatic system, although other factors may cause protons to stray into this region.

## Mesomeric effects

In unsaturated systems, electron density may also be influenced by the presence of groups which induce mesomeric effects. For example, compare the influence of the +M effect of the methoxy group with the −M effect of

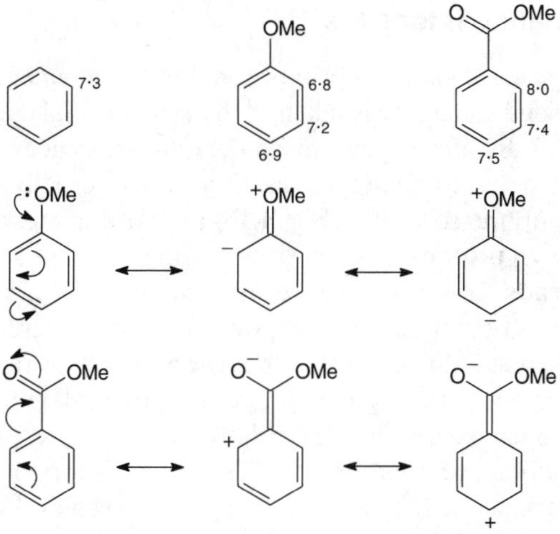

**Fig. 4.14** Mesomeric effects influence electron distribution and hence chemical shifts. – indicates an increase in electron density, and hence increased shielding whereas + indicates a decrease in density and hence deshielding of the protons.

the ester group in Fig. 4.14. Clearly, these effects can be operative over many bonds, in contrast to the shorter-range inductive effects, which would also be present.

## Hydrogen bonding

Hydrogen bonds, usually involving OH, NH or SH groups, have an electron withdrawing effect on the proton involved and may move such protons to high frequencies by many ppm. They may occur intramolecularly or intermolecularly. The shift of any hydrogen bonded proton tends to be hard to predict with any accuracy as it generally represents a population-averaged shift between that of the hydrogen bonded state and of the non-hydrogen bonded state (see Section 4.6 for more discussions on averaging effects in NMR spectra). Shifts due to intermolecular hydrogen bonds are therefore concentration dependent. For example, carboxylic acids often occur in solutions as hydrogen bonded dimers, so that the acidic protons resonate in the 11–14 ppm region (due to a combination of inductive and hydrogen bonding effects), but shift to lower frequency as the solution becomes more dilute. Temperature also has a large influence on the shift of hydrogen bonded protons, with increases in temperature generally causing shifts to lower frequency, due to the breaking of these bonds.

In addition to the effects described above, it should be noted that the solvent itself can also influence chemical shift values. Indeed, a change of solvent can often be a valuable tool in aiding the interpretation of a spectrum, as resonances that are coincident in one solvent may be well dispersed in another. Shift differences are often observed when changing from a relatively non-polar solvent to a more polar one, or to an aromatic solvent. Changing from chloroform to benzene can often have a dramatic effect (Fig. 4.15); solvation of the solute by benzene produces these changes because of the magnetic anisotropy associated with the aromatic ring, which is clearly not present with chloroform.

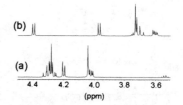

**Fig. 4.15** Changing solvent can have a dramatic effect on the distribution of chemical shifts; (a) was recorded in chlorofom and (b) in benzene.

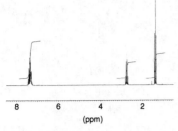

**Fig. 4.16** Integration of resonances may be used to determine the relative number of protons present in different chemical environments. For ethyl benzene the ratio is 5:2:3.

Integrals only represent the **relative ratio** of protons in each environment, so don't be caught out by symmetrical molecules. For example, the integrals for diethyl ketone would be 2:3, although they actually represent 10 protons.

Unlike the information provided by all other spectroscopic techniques, where an absorption tells us only about the **absorbing species itself**, spin-spin coupling provides information about the **neighbours** of the absorbing species.

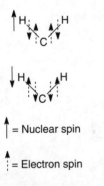

↑ = Nuclear spin

↑ = Electron spin

**Fig. 4.17** The orientation of one spin can be relayed through the intervening bonding network *via* electron spins to a neighbouring nucleus. This is know as **scalar coupling**.

## 4.3   Resonance intensites

The intensity of a resonance is proportional to the number of nuclei giving rise to it, provided some care is taken in the experimental conditions used to acquire the data. By intensity, we mean the total area under a peak, which is readily obtained by computer integration and is usually presented as a separate trace on the spectrum (Fig. 4.16) and/or a numerical value. Thus, Fig. 4.16 shows proton resonances with integral ratios of 5:2:3, and provides a ready count of the relative number of nuclei in a given environment. Although computers can readily report integrals to many decimal places, it should be stressed that the accuracy of the measurements *on routine spectra* is generally low, for reasons that relate to the behaviour of the nuclei once excited by the radiofrequency pulse, which are more appropriately discussed in Section 5.2. The values reported may often be in error by as much as *ca.* 10% lower than their true values. Furthermore, the degree of inaccuracy will not necessarily be the same for all resonances in a spectrum, so an exact comparison of integrals is futile. However, if you simply wish to determine whether a resonance represents 1, 2 or 3 protons within a molecule, the level of accuracy is usually sufficient to answer this. Use of integrals for exact quantitative determinations (such as determining the ratio of two compounds in solution) is, however, possible, although special procedures are required.

## 4.4   Spin-spin coupling

Coupling between nuclear spins gives rise to splitting of resonance lines, providing the chemist with evidence for chemical bonding within a molecule. This information can be obtained because the fine structure seen within resonances arises from a **through bond** effect, and so can be easily related to the common perception of chemical bonding. This information is fundamentally different from that obtained from UV or IR spectroscopy in that these techniques only give evidence for the **presence** of functional groups whereas spin coupling tells us about how nuclei **interact** with each other. How does this effect arise?

### The origin of spin-spin coupling

The magnetic field of a nucleus in a chemical bond will directly influence both the nuclei immediately surrounding it and the electrons in the bond. As we have seen above, the spin-$\frac{1}{2}$ nucleus can adopt two possible orientations in the applied field. Each orientation will have a differing influence on the electrons in the bond as the electron spins will favour an anti-parallel orientation relative to the nuclear spin. These electrons relay the effect of the different nuclear spin states according to the electron spin pairing rules of the Pauli exclusion principle and of Hund's rules (Fig. 4.17). Hence, a nucleus is able to sense the possible orientations of a neighbouring nucleus when there is a suitable bonding network between them, and because the effect is operative only through bonds, it is referred to as indirect or *scalar coupling*

The *direct* influence of one nucleus on another is easier to imagine as it is somewhat analogous to the effect that two bar magnets would have on one another. Any nuclear spin will experience a different magnetic environment for each of the two possible (α and β) states of its neighbour, according to whether the neighbouring field enforces or opposes the static field (Fig. 4.18). However, in solution the molecule tumbles rapidly, so that the orientation between the two nuclei and the static field is constantly altering and the fields they experience due to each other fluctuate as the molecule rotates. Because of the motion, the overall effect of the two states on a neighbouring nucleus averages to zero, so that the *direct interaction between nuclei produces no splittings in a spectrum* (this is not the case for spectra recorded in the solid state, where such rapid molecular motion does not exist to remove the effect). The direct coupling is known as *dipolar coupling* as the spins are considered to be magnetic dipoles (North and South). Although it is not seen in solution spectra, its presence can be detected with suitable experiments, and it provides the mechanism for one of the most useful NMR phenomenon available; the nuclear Overhauser effect (Section 5.3). Scalar coupling, however, is not averaged to zero by molecular motion, because of the through-bond dependence, and hence the following considerations will relate to scalar coupling only.

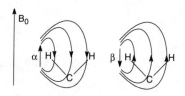

**Fig. 4.18** The direct magnetic influence of one spin on another is known as *dipolar coupling.*

## The influence of scalar coupling

Consider a molecule containing two chemically distinct coupled protons, designated A and X. So far we have seen that it is possible for a nuclear spin, say A, to sense the state of the neighbouring nucleus, X, to which it is coupled, *via* the bonding electrons. The two possible orientations of X, α, and β, will result in two slightly different local fields at A. Now, the sample as a whole will contain very many molecules and, as the population differences in NMR are so small (1 part in $10^4$ at most), we can say that there will be approximately an equal number of X nuclei in the α state as in the β state. Thus, effectively, half the transitions for all the A nuclei in the sample will now occur at one frequency, resulting from the Xα state, with the other half resonating at a different frequency resulting from the Xβ state. The A resonance will, therefore, constitute two lines; it will be a doublet. Similar arguments will apply to the X nuclei, so they will also give a doublet (Fig. 4.19)

The frequency difference between the two lines of the doublet is known as the *coupling constant*, *J*. As the coupling is dependent on nuclear and electron spin pairings *within* a molecule, the coupling constant is independent of the external field, so is always measured in hertz. The effect is also symmetrical, so that the coupling constant measured at A will be equal to that measured at X ($J_{AX} = J_{XA}$) and this equality provides a means of identifying nuclei that are coupled together and hence contained in a bonding pathway. In due course we shall see how this information may be related to structural fragments within a molecule.

Coupling constants possess *sign* as well as *magnitude*. The sign of *J* is not apparent in routine spectra, and may only be determined from specially designed experiments or, in some cases, from complex analysis of multiplet structures, so generally it is only the absolute value that is considered. It is important to be aware of the existence of sign because substituent effects within a molecule that increase coupling (make it more positive) may actually reduce the *measured value* of *J* if the coupling constant has negative sign. Coupling constants are defined as being

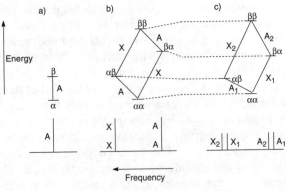

**Fig. 4.19** Energy level diagrams (above) and associated schematic spectra (below) for uncoupled and coupled spins. (a) A single spin has only a single transition. (b) Two spins without coupling (J = 0) show a single resonance for each spin as the two possible transitions of each are degenerate. (c) Coupling between the spin (J > 0) removes the degeneracy and each spin now has two transitions of differing energy; each produces a doublet.

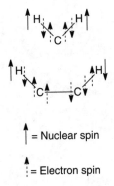

↑ = Nuclear spin

↑ = Electron spin

**Fig. 4.20** Favoured nuclear and electron spin pairing in two-bond and three-bond coupling pathways.

The terms **geminal** and **vicinal** come from the Latin *geminus*, meaning twin, and *vicinus*, meaning neighbour.

positive if the two coupled spins have an anti-parallel orientation in a lower energy level than that of the uncoupled spins, and negative when parallel spins are favoured. Thus, the spin system in Fig. 4.19 (the AX system described above) represents a positive coupling constant, as the anti-parallel spins, $\alpha\beta$, have lower energy when coupled (Fig. 4.19c) than when they are not (Fig. 4.19b). The favoured interactions for two protons coupled over two and three bonds are shown in Fig. 4.20 and it follows that couplings over three bonds are usually positive whereas those over two bonds are usually negative.

So far we have given consideration to how coupling between spins arises through a suitable bonding network and how this affects resonances, but what consitutes a suitable bonding network? In proton spectroscopy, $^{1}H$–$^{1}H$ couplings are most commonly seen to occur over only two or three bonds and are referred to as **geminal** and **vicinal** couplings respectively. In favourable cases, as we shall see later, couplings over more bonds may occur and these are generally referred to as **long-range** couplings. The full range over which $^{1}H$–$^{1}H$ coupling constants have been observed is $-24$ to $+40$ Hz and discussions as to what influences the magnitude of *J* are presented in a later section. Note also that the nature of nuclei intermediate on a coupling path does not interfere with the coupling mechanism, so that coupling between two protons (or indeed any nuclei) may occur even though other nuclei on the bonding pathway may have zero spin.

### Interpreting coupling patterns

The measurement of chemical shifts and coupling constants directly from spectra of coupled nuclei is strictly only possible when the spectra satisfy the **first-order approximation** (also known as the weak coupling approximation).

This applies when the chemical shift difference between two coupled resonances is large compared with their coupling constant (what is meant by large is discussed later). For many coupled resonances, the first-order model is valid, or nearly so, allowing direct analysis of spectra; the following guidelines for interpreting patterns assume this to be the case.

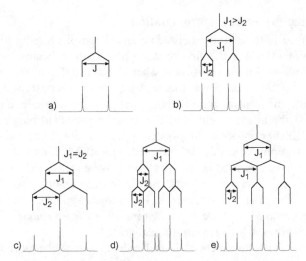

**Fig. 4.21** Simple multiplet patterns arising from scalar coupling. (a) doublet, (b) double doublet, (c) triplet, (d, e) double triplets (see text also).

We have seen above that coupling to a single nucleus of spin-$\frac{1}{2}$ splits a resonance into two lines of equal intensity, separated by the coupling constant, $J$. Coupling to further nuclei will simply repeat this line-splitting process according to the subsequent coupling constants, so that coupling to many nuclei may be analysed as a number of discrete stages. Thus, coupling to two different protons will reduce the resonance to four lines, three protons to eight lines and so on. The total area of a resonance remains constant (the integral must reflect the relative number of protons giving rise to the resonance) so that the introduction of coupling reduces the intensity of each line. The scheme in Fig. 4.21 presents the analysis of some simple coupling patterns.

Although the exact details of how a multiplet will appear will depend on the magnitudes of $J$ involved, the chemical shift of the resonance always occurs at the centre of the multiplet and the magnitudes of the coupling constants can be measured directly from the line splittings.

In some instances the magnitude of coupling to two or more protons may be the same, either because the protons are indistinguishable, those in a methyl group for example, or simply by a matter of chance. In such cases the number of lines will not follow the simple rule outlined above as certain lines will become coincident (Fig. 4.21 c, d, and e), although it is possible to predict the relative line intensities by the total number of 'branches' contributing to each line. Coupling to '$n$' equivalent spin-$\frac{1}{2}$ nuclei produces line intensities that follow the coefficients of the binomial expansion of $(1 + x)^n$ which are conveniently displayed as *Pascal's triangle* (Fig. 4.22). These patterns are evident for the ethyl group of ethyl *p*-tolylacetate (Fig. 4.5), which displays the quartet of the $CH_2$ group and the triplet of the $CH_3$ group. Note that *protons that are equivalent show no coupling amongst themselves*, so that, for example, the three protons of the methyl in an ethyl group show couplings only to the neighbouring methylene (more detailed discussions of what is meant by 'equivalent' in an NMR sense, as opposed to the chemical sense, are presented later).

| $n$ | | |
|---|---|---|
| 0 | 1 | Singlet |
| 1 | 1 1 | Doublet |
| 2 | 1 2 1 | Triplet |
| 3 | 1 3 3 1 | Quartet |
| 4 | 1 4 6 4 1 | Quintet |
| 5 | 1 5 10 10 5 1 | Sextet |
| 6 | 1 6 15 20 15 6 1 | Heptet |

**Fig. 4.22** Pascal's triangle. Coupling to $n$ equivalent spin-$\frac{1}{2}$ nuclei produces $n + 1$ lines, the relative intensities of which are given by the triangle.

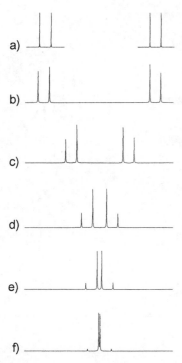

**Fig. 4.23** Simulations of a coupled two spin system showing the effect of reducing the shift difference between the resonances. The $\Delta\delta/J$ ratios are a) $\infty$ (the AX system) b) 10 c) 5 d) 2 e) 1 and f) 0.5

## Failure of the first-order approximation

As the chemical shift difference between coupled protons is reduced, there comes a point at which the first-order approximation begins to break down. As a rule of thumb, this occurs when the shift difference (measured in hertz) is less than ten times the relevant coupling constant. As the chemical shift differences are in hertz, they will be dependent on the magnitude of the applied field, and the use of higher-field magnets will increase the likelihood of the first-order approximation being valid, and so aid spectroscopic analysis. In the preceding discussions it has been assumed that lines observed in the spectrum may be attributed to a single energy transition, or in other words, the flip of only a single spin. This is not the case when the first-order model no longer applies as the energy states become 'mixed' and we say the spins are ***strongly coupled***. In this case, three significant features can be recognized:

(a)  The centres of multiplets no longer reflect the true chemical shift of the resonance.

(b)  Line splittings often do not represent coupling constants.

(c)  Line intensities become distorted.

The extraction of $\delta$ and $J$ can be made by detailed analysis of the fine structure, or more commonly nowadays, with the use of computers to simulate the spectra. Whilst these steps are not usually pursued in routine applications, it is important to be able to recognise when the first-order model fails, so that incorrect values of $\delta$ and $J$ are not recorded. When analysing spectra, it is common practice to label each proton under consideration with a letter of the alphabet, and the further apart the protons in the spectrum, the further apart are the letters chosen. Thus, in the simple two spin system considered above, protons were designated A and X, to indicate a large shift difference, consistent with adherence to the first-order model. Fig. 4.23 shows what happens when the shift difference, $\Delta\delta$, between the two protons is reduced. Initially the protons change from an ***AX*** system to an ***AB*** system system until eventually they become equivalent (an $A_2$ system) and show only a single line.

In the AB system, the intensities of the inner lines increase at the expense of the outer lines; this so-called ***roofing effect*** can be very helpful in indicating coupling partners, or at least indicating where the partner may be. For this simple system, the separation between the lines still represents the true value of J. However, the chemical shift is no longer at the midpoint of the 'doublet' but is weighted toward the inner (more intense) line until, for the $A_2$ system, the chemical shift corresponds with the shift of the single line. The AB system is often seen for lone $CH_2$ groups (with no other coupling partners) for which the two protons are chemically inequivalent, such as when in a rigid ring system or adjacent to a stereogenic centre (see Section 4.5).

Another commonly observed pattern is that of a three spin system. If the shift difference between all three protons is very large relative to the coupling between them, then the spin system would be designated an ***AMX*** system and can be analysed as a first-order spectrum as described in the previous section. However, it is often the case that two of the protons will have similar shifts, whilst the third is distant from these, such as in $RCH_2CHXY$ or $XCH=CH_2$. In this case the spin system is designated

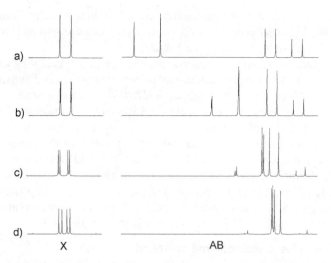

**Fig. 4.24** Simulated ABX spectra showing the effects of a reduction in the shift difference between the A and B parts. The simulations used $J_{AB} = 15$, $J_{AX} = 6$ and $J_{BX} = 0$ Hz with $\Delta\delta/J_{AB}$ ratios of a) 5 b) 2 c) 1 and d) 0.5.

***ABX***, and strong coupling effects are observed. For example, the spectra in Fig. 4.24 show the effect when the shift difference between spins A and B is progressively reduced. The simple first order rules would predict that the X part of the spectrum would only be a doublet, as it has coupling to proton A only, but obviously this is not the case as the X resonance has additional fine structure. This phenomenon is referred to as ***virtual coupling*** to distinguish it from first-order scalar coupling.

Clearly, one should be aware of such effects when analysing spectra so as not to interpret splittings arising from strong-coupling as being due to additional coupling to other protons. Notice that some of the outer lines of the AB part have very low intensities in some cases and it is not uncommon for these to be lost in noise or hidden by other resonances in real spectra, further complicating analysis.

***ABC*** systems in which the shifts of all three spins are close together, produce very complex spectra (Fig. 4.25), the appearance of which will be very sensitive to the chemical shifts and coupling constants involved, and are best analysed with the aid of a suitable spin-simulation program.

Before we consider other instances in which the first-order approximation is not valid, it is necessary to define what is meant by equivalence in NMR terms. This may take two forms; ***chemical equivalence*** or ***magnetic equivalence***. For two nuclei to be chemically equivalent they must resonate at the same chemical shift and have the same chemical properties, for example the 2 and 6 protons of a 1,4 disubstituted benzene ring are chemically equivalent due to symmetry (Fig. 4.26). For nuclei to be magnetically equivalent they must *also* both demonstrate the same couplings to all other nuclei, which does not apply to the 2 and 6 protons in the above example. Thus, proton $H^2$ has coupling to $H^3$, $H^6$ and $H^5$ with $J_{23} > J_{26} > J_{25}$, whereas $H^6$ (the chemically equivalent partner) couples to the same protons but has $J_{56} > J_{26} > J_{36}$ and, by symmetry, $J_{23} = J_{56}$ and $J_{25} = J_{36}$. Clearly the coupling constants of protons 2 and 6 to, say, proton 3 are not the same, so 2 and 6 cannot be magnetically equivalent. In contrast,

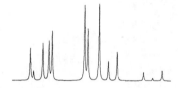

**Fig. 4.25** A simulated ABC spectrum with $J_{AB} = 15$, $J_{AC} = 6$ and $J_{BC} = 0$ Hz, $\Delta\delta_{AB}/J_{AB} = 1$ and $\Delta\delta_{AC}/J_{AC} = 4$.

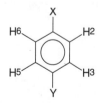

**Fig. 4.26** The 2 and 6 or 3 and 5 protons in a 1,4 disubstituted benzene ring are ***chemically equivalent*** but not ***magnetically equivalent*** (see text).

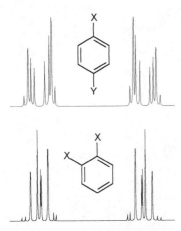

**Fig. 4.27** The presence of *chemical* but not *magnetic* equivalence in molecules produce spectra that are more complex than may be anticipated. These AA′BB′ spectra display considerably more fine structure than would be anticipated for a simple $A_2B_2$ system. The chemical shift degeneracy means that use of higher-field spectrometers *does not* alleviate these complications.

the protons in a freely rotating methyl group are magnetically equivalent because they all experience the same chemical environment and will have the same couplings to any neighbouring nuclei.

Nuclei that are chemically equivalent but not magnetically equivalent are indicated by the use of an additional prime, so that the 1,4 disubstituted benzene described above becomes an AA′XX′ or AA′BB′ system.

Generally, systems which possess chemical but not magnetic equivalence are more complex than may be anticipated because they do not follow the first-order approximation. The AA′BB′ spectra (Fig. 4.27) which are commonly observed for 1,2 and 1,4 disubstituted benzenes demonstrate the complexity that can arise in these circumstances. They are greatly different from the simple $A_2X_2$ or $A_2B_2$ spectra that may have been anticipated for these systems, and are often mistakenly assigned as 'AB quartets', which clearly they are not (compare Fig. 4.23).

## $^1$H–$^1$H coupling constants and chemical structure

From the previous discussions it may appear that any single proton in a molecule would be able to couple with all other protons within the same molecule *via* the various interconnecting bonding pathways. Fortunately for the chemist, this is not the case as the degree of coupling falls rapidly as the number of intervening bonds increases, so that coupling is only usually observed over two or three bonds. Of these, the three bond (vicinal) coupling is of most use in structure determination, as it offers a means of identifying neighbouring heteroatom (usually carbon) groups, to which the protons are attached, and hence piecing together the skeleton of the molecule. Spin coupling between protons separated by four or more bonds can be observed in favourable cases and is termed long-range coupling for obvious reasons. Values of typical $^1$H–$^1$H coupling constants are presented in Fig. 4.28 and the influence of various structural features on these values is discussed in the following paragraphs.

*Geminal coupling.* The full range of geminal coupling constants, $^2J$, occurs over −20 to +40 Hz, but it is most common to encounter values around −10 to −15 Hz for saturated $CH_2$ groups (typically −12 Hz in the absence of electronegative substituents) and −3 to +3 Hz (typically −2 Hz) for terminal alkenes. Hybridization of the carbon atom has a large influence on the magnitude of $^2J$; a change from sp³ to sp² and the

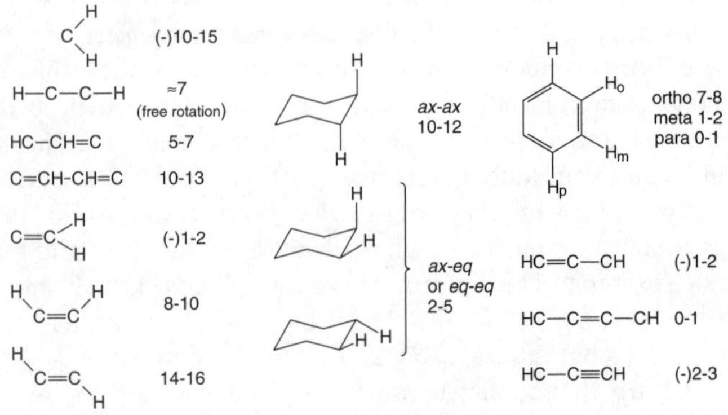

**Fig. 4.28** Examples of some typical $^1$H–$^1$H coupling constants (Hz).

associated increase in H–C–H bond angle causes an increase in the size of $J$, that is, it becomes more positive. The addition of electronegative groups to the carbon will also make $J$ more positive, and the effect of further substituents is approximately additive (Fig. 4.29). Remember that only the absolute value of a coupling constant is represented in the spectrum, so that the value of $J$ measured will actually *decrease* when $J$ becomes more positive, if it has negative sign.

The presence of electronegative groups on the carbon atom adjacent to the methylene of interest, or unsaturation at this adjacent carbon, will cause $J$ to become more negative.

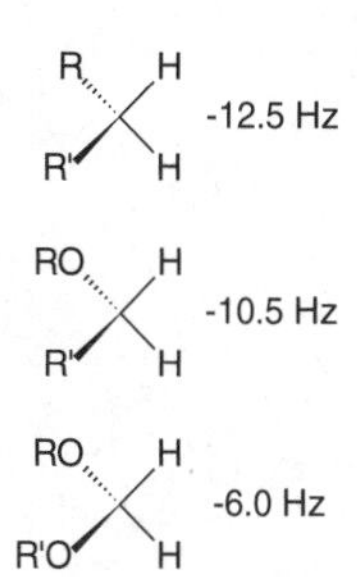

*Vicinal coupling.* Vicinal coupling constants, $^3J$, are always positive in sign and may take values of up to 14 Hz in saturated or 18 Hz in unsaturated systems, although the average value of $^3J$ across a freely rotating C–C bond is around 7 Hz. Four major influences on the magnitude of vicinal coupling constants can be identified and are considered further below, they are
(a) substituent effects,
(b) bond lengths,
(c) bond valence angles,
(d) dihedral angles.

The introduction of electronegative groups on the HCCH pathway tends to reduce the magnitude of $^3J$ in both saturated and unsaturated systems. The effect is usually quite small, however ($< 1$ Hz), but has a maximum influence when the electronegative group is *trans* to one of the protons, as can be observed in cyclohexane systems (Fig. 4.30).

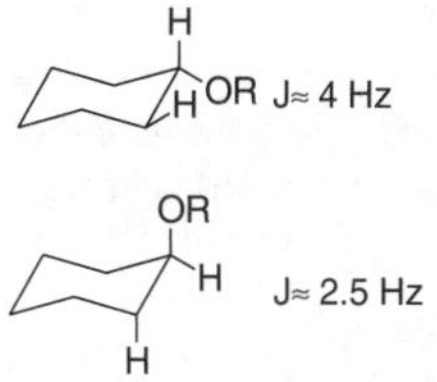

Vicinal coupling constants are very sensitive to the C–C bond length with an increase in length causing a decrease in $J$. For example, the C–C bonds in benzene possess less double bond character than those of alkenes, resulting in a slightly longer C–C bond length in the former, which is manifested in the different $^3J$ values for benzene and for Z-alkenes of *ca.* 8 and 10 Hz respectively.

An increase in the valence angle, $\theta$, causes a reduction in the value of $^3J$, as can be seen for the protons in cyclic Z-alkenes of various ring sizes (Fig. 4.31).

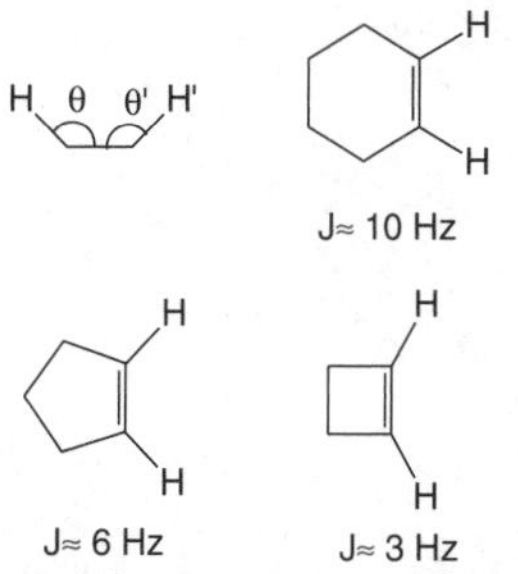

The variation of $^3J$ with the dihedral angle is arguably the most useful of the four categories discussed here as it provides a means of conformational analysis and often proves useful in addressing problems of stereochemistry. The dependence of vicinal coupling constants on the dihedral angle, $\phi$, is represented by the Karplus Curve, as represented in Fig. 4.32, which was first predicted theoretically, and has since been confirmed by experimental observation. The theoretical curve follows the relation

$$^3J = 4.22 - 0.5\cos\phi + 4.5\cos 2\phi.$$

The actual values of $^3J$ encountered are, of course, also dependent on substituent effects, although the general shape of the curve is always reproduced, and $^3J_{180}$ always has a larger value than $^3J_0$. This latter point explains the differences that are observed for $^3J$ values in Z and E alkenes; the former have $\phi = 0°$ and have $^3J$ in the 8–11 Hz range whereas the latter have $\phi = 180°$ and have typical $^3J$ values of 12 -16 Hz (the large

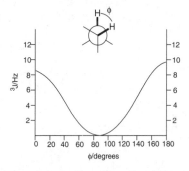

**Fig. 4.29** Substitution of electronegative groups causes the $^2J$ coupling constant to become more positive, although the apparent splitting may *decrease* as $J$ may be negative.

**Fig. 4.30** Electronegative substituents on the coupling pathway tend to decrease $^3J$.

**Fig. 4.31** Increases in the bond valence angle tend to decrease $^3J$.

**Fig. 4.32** Variation of $^3J$ with bond dihedral angle is represented by the Karplus curve.

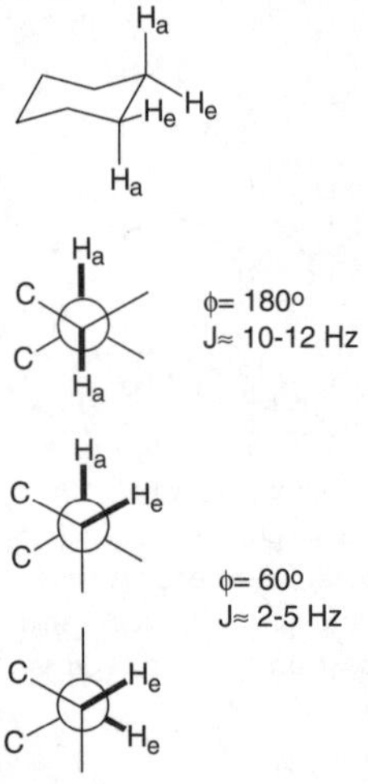

**Fig. 4.33**  The variation of $^3J$ with dihedral angle can be useful in the assignment and conformational analysis of cyclohexanes.

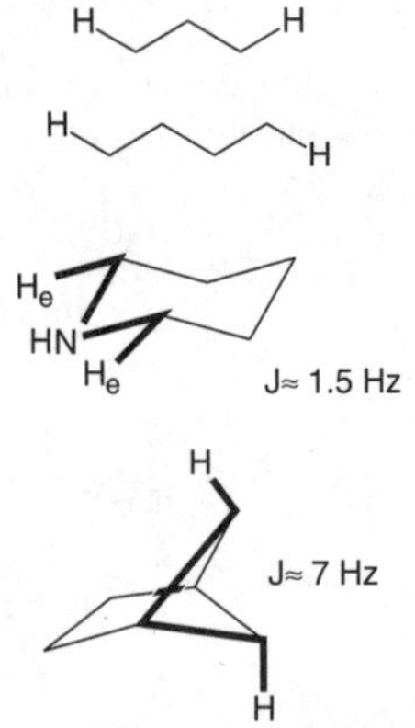

**Fig. 4.34**  Favoured 'zig-zag' coupling pathways for observable long-range coupling in saturated systems.

magnitudes of these couplings, relative to those encountered in unsaturated systems, arise from the shorter $C=C$ double bonds).

Similarly, $^3J$ values for *axial-axial* protons ($\phi = 180°$) are larger than those of *axial-equatorial* or *equatorial-equatorial* partners ($\phi = 60°$) in cyclohexanes, as can be appreciated from Fig. 4.33. This provides a very useful criterion in the study of the conformation of these ring systems. Finally, notice that the value of $^3J$ can become vanishingly small for $\phi$ angles close to 90°, so that the lack of an observable coupling between two protons need not preclude them from being only three bonds apart.

*Long-range couplings.*    Long-range couplings are those that operate over more than three bonds. They are most commonly seen to occur over 4 or 5 bonds and have $^nJ \leq 3$ Hz. In saturated systems, four-bond couplings may be observed when the protons sit in a 'zig-zag' relationship (the so-called *w*-coupling, Fig. 4.34), but are rarely seen otherwise. The magnitude of the four-bond coupling may also be increased by ring strain, such as in the bicyclo[2.1.1] hexane in Fig. 4.34.

The appearance of long-range coupling is most likely when the coupling pathway includes unsaturation, and is again favoured by a 'zig-zag' relationship. Allylic couplings ($H–C=C–C–H$) of 1–2 Hz are widespread, and coupling over 5 or more bonds may be detected in conjugated systems; although these may not always be resolved they may be manifested as line-broadening. Four bond couplings in aromatic systems are commonly encountered (typically 1–3 Hz), and result in spectra that are more complex than may first be supposed, due to the introduction of magnetic inequivalence (see above).

## Spin decoupling

Previous discussions have told us that it is possible to identify mutually coupled protons by virtue of them having the same coupling constant with one another. Whilst this is indeed the case it is often found in practice that a simple analysis of the spectrum does not provide an unambiguous assignment of coupled spins. For example, two coupled protons may exhibit a mutual coupling constant of 7 Hz suggesting them to be partners, but what if a third proton also coincidentally exhibited a coupling constant of 7 Hz to its partner? It is then not possible to say which of these protons are mutually coupled. Difficulties may also occur when a proton resonates in a crowded region of the spectrum so that its multiplicity and coupling constants cannot be determined directly. One technique that aims to establish connectivities unambiguously is that of *spin decoupling* in which radiofrequency irradiation is applied at the frequency of one resonance during the time the spectrum is being recorded. This induces rapid transitions between the $\alpha$ and $\beta$ states for the irradiated proton so that the coupled partner senses only an average of the two spin states. Consequently, the coupled partner resonates at only a single frequency (in the absence of further coupling to other protons in the molecule) and the coupling between the protons appears to have been removed; they are said to have been **decoupled** from each other. Only couplings to the irradiated proton will be removed by this process so differences in spectra recorded with and without the irradiation will indicate the shifts of protons that are spin coupled to the irradiated proton (Fig. 4.35). In complex spectra the two experiments are often subtracted so that only differences between the two remain, thus making identification of coupled protons easier, and is termed ***decoupling difference spectroscopy***. Because we are both irradiating and observing protons, the technique is referred to as ***homonuclear decoupling***

## 4.5   Chirality and NMR

In many instances the two protons of a $-CH_2-$ group are seen to resonate at different frequencies and to show a mutual coupling, that is, they are chemically inequivalent, and are referred to as ***diastereotopic*** protons. This inequivalence occurs when the methylene group is adjacent to an asymmetric centre, such as the fragment $WCH_2-CXYZ$, and is not lost by rapid rotation about the C–C bond. The reason for this can be understood from considering the Newman projections for the three energetically favoured rotamers, but the argument applies equally to any rotamer (Fig. 4.36).

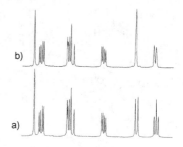

**Fig. 4.35**   Homonuclear spin decoupling reveals the coupling partners of the irradiated proton (not shown). (a) Standard $^1H$ spectrum, (b) decoupled spectrum.

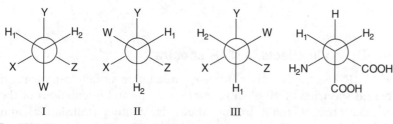

**Fig. 4.36**   Newman projections of the three energetically favoured rotamers (see discussions in text).

From inspection it may be seen that, no matter what rotameric form, the chemical environments for the two protons are never mutually interchanged. For example, the environment of $H_1$ in rotamer I is not the same as that of $H_2$ in rotamer III, and so on, thus giving rise to chemical inequivalence. As an example, the diastereotopic methylene protons of aspartic acid are part of an ABX, not an $A_2X$ system, (Fig. 4.37) with each proton coupled to its geminal partner and to the proton on the adjacent chiral carbon. Replacing the W group with a proton will remove the differences between rotamers, so that all protons will experience the same average environment as the methyl group rotates, and hence will be equivalent.

Inequivalence is not only limited to the protons of methylene groups, however, so that a pair of methyl groups may also be diastereotopic, for example, the two methyl groups of valine $(H_2NCH(CH(CH_3)_2)COOH)$ are inequivalent and resonate at 0.97 and 0.94 ppm.

The idea of diastereotopic groups can be made more general, as a stereogenic centre is not always required for inequivalence. In fact, methylene protons adjacent to *any group that lacks symmetry* will be diastereotopic. Thus, in the group $R-CH_2-CXY-CH_2-R$, which is prochiral, the $-CXY-CH_2R$ group itself has no symmetry so that the methylene protons remain inequivalent. Groups that are equivalent because they share a mirror image relationship with one another are referred to as being ***enantiotopic***.

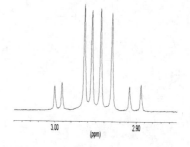

**Fig. 4.37**   The $CH_2$ group of aspartic acid displays resonances characteristic of the AB part of an ABX system. The protons are inequivalent and are said to be ***diastereotopic***.

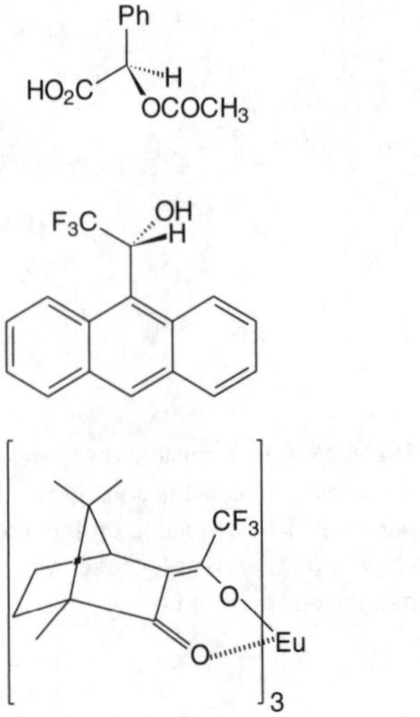

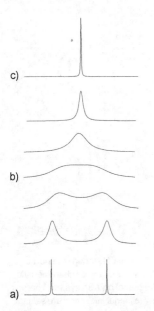

**Fig. 4.38**   Example chiral shift reagents. They all contain functional groups capable of hydrogen bonding with the solute molecules of interest, thus forming diastereomeric complexes.

**Fig. 4.39**   Simulated spectra for the dynamic exchange of two equally populated sites. (a) slow-exchange regime, (b) coalescence point, (c) fast exchange regime.

In some instances, peaks in the intermediate exchange regime can become so broad they are lost in the baseline and effectively disappear.

Now let us briefly consider the analysis of enantiomers by NMR. Whilst the NMR spectra of diastereoisomers may be different, those of enantiomers are indistinguishable (radiofrequency is an achiral probe). However, it is often desirable to be able to assess the relative ratios of two enantiomers in solution, and this is made possible by the use of **chiral shift reagents**. Such reagents are single, pure enantiomers of a molecule that will interact with the species to be analysed when mixed in solution, forming a **diastereomeric complex**. Some example reagents are shown in Fig. 4.38. The complex formation is brought about by the interaction of functional groups, and the shift differences of the diastereomeric complexes arise from the anisotropy of the shift reagent. The complexes now have spectra that can be distinguished, and the relative ratios of the enantiomers is determined from integration of suitable resonances. The ratio is usually expressed as the **enantiomeric excess** (*ee*), where

$$ee = \frac{I_A - I_B}{I_A + I_B} \times 100\%$$

and $I_A$ and $I_B$ represent the integrals of the two complexes arising from enantiomers A and B. Clearly, the *ee* is zero for a racemic mixture, as $I_A = I_B$.

## 4.6   Dynamic effects in NMR spectra

The NMR spectra of most molecules in solution usually represent an averaged spectrum of all possible conformations and orientations of the molecule, brought about by fast molecular motion (translations and rotations). In this section we consider what happens when the motion can no longer be consider fast, and indeed what defines 'fast' in the context of NMR analysis.

A commonly encountered situation which may occur on a timescale that gives rise to dynamic effects in NMR spectra is that of hindered rotation about a bond. Consider a proton that sits in a molecule that may take up two possible conformations of equal population. The proton can be expected to resonate at different frequencies in each conformer, let the difference between these two be $\delta v$ Hz. If we wished to characterize the two resonant frequencies and so observe each conformer separately, the Heisenberg uncertainty principle tells us that each conformer must have a lifetime of at least $1/(2\pi\delta v)$ seconds. If the lifetimes are less than this the chemical shifts for the proton in each environment cannot be determined, and only a single resonance is observed at an average chemical shift. Thus, if the rate constant, $k$, for an exchange process is much greater than $2\pi\delta v$ ($s^{-1}$), the process is fast on the NMR timescale (or more specifically the chemical shift timescale) and an averaged spectrum is obtained. If the rate constant is much less than $2\pi\delta v$, the process is slow and separate resonances are observed for the two conformers (Fig. 4.39). In between these two extremes the exchange is said to be intermediate; resonances become broad and the two separate peaks become one at the coalescence temperature, before forming the single time-averaged peak. Rate constants within the range of approximately $10^{-1}$ to $10^3$ $s^{-1}$ can be expected to produce directly observable changes to the NMR spectrum. Such effects are often observed in the spectra of amides in which rotation about the amide bond is restricted due to its partial double-bond character (Fig. 4.40).

Studies of changes in lineshapes with temperature can lead to the determination of rate constants for a number of dynamic processes, but are beyond the scope of this text. For the simple case of two-site exchange with equal site populations the rate constant at the coalescence point, $k_c$, is given by

$$k_c = \frac{\pi \delta v}{\sqrt{2}} \; s^{-1}$$

and the free energy of activation, $\Delta G^{\ddagger}$, for the process at this coalescence temperature, $T_c$ (in K), may be derived from the Eyring equation as

$$\Delta G^{\ddagger} = 8.31 T_c \, [22.96 + \ln(T_c/\delta v)] \; J \; mol^{-1}$$

for which the value of $\delta v$ must be determined from the shift difference between the resonances in the slow-exchange regime.

Because the exchange timescales are dependent on the chemical shift difference *in hertz* between the exchanging resonances, the appearance of the spectra of dynamic systems may vary with the applied field strength. For example, shift differences are larger at higher field strengths, so that exchange processes may appear slow whereas on a lower-field instrument they may be in the intermediate or even fast regime; use of a higher field strength has a similar effect on the appearance of spectra as would cooling the sample (although in the latter case it is the actual exchange rates that alter). Likewise, observing another nucleus may provide very different results from those seen in proton spectra, because the shift differences are unlikely be the same as they were in the proton spectrum.

Resonance broadening may also come about by intermolecular exchange processes, for example the resonances of exchangeable protons such as OH, NH or SH are often broad and show no resolved couplings, usually due to acid-, but sometimes base-, catalysed exchange. In a moiety such as CH–OH you might expect to observe splitting of the CH resonance due to the neighbouring hydroxyl proton, giving rise to a doublet in the absence of further couplings. The splitting arises, as described in Section 4.4, because the OH may take up two possible orientations, $\alpha$ and $\beta$, causing CH protons to resonate at two frequencies. Suppose the hydroxyl proton initially has the $\alpha$ arrangement in a particular molecule, but is then exchanged for another proton. The incoming proton may have the same ($\alpha$) arrangement, in which the CH resonant frequency is unchanged, or it may have the opposite ($\beta$) arrangement, in which case the CH proton now resonates as the other half of the doublet. If the exchange of the hydroxyl proton is sufficiently fast, the two lines of the doublet will never been seen, and a single averaged line is observed. The CH–OH pair are said to be *exchange decoupled*. If the exchange rate was very slow, both transitions could be observed and the coupling would be apparent on the CH and OH resonances (Fig. 4.41).

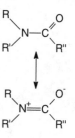

**Fig. 4.40**   Slow rotation about amide bonds is often observed due to the partial double-bond character.

Altering experimental conditions (temperature or field strength) to move into the slow-exchange regime is often said to 'freeze out' the individual spectra of interconverting conformers.

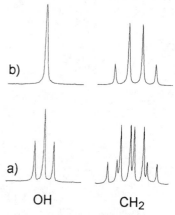

b)

a)

OH                    CH$_2$

**Fig. 4.41**   Partial $^1$H spectra of ethanol in (a) the absence and (b) the presence of acid. Coupling between the methylene and the hydroxyl protons is not seen in (b) due to *exchange decoupling*.

A simple test for the presence of exchangeable protons, often referred to as the 'D$_2$O shake', is to add two drops of D$_2$O to the NMR tube, mix, leave to settle and re-record the spectrum. Exchangeable protons should disappear from the spectrum since they will have been replaced by deuterons.

## 4.7   Exercises

1.   Assign the $^1$H spectrum of ethyl *p*-tolylacetate shown in Fig. 4.5 using the chemical shifts and coupling patterns only.

2.   Assign the spectra below to the three structural fragments shown by analysing the coupling fine structure in these spectra.

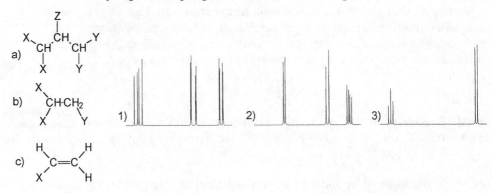

3.   An unknown molecule gave the $^1H$ spectrum shown below. Based on this, and the knowledge that the molecular formula is $C_5H_{10}O$, suggest a structure for the unknown.

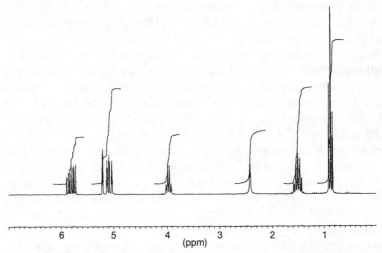

4.   At room temperature, the 200 MHz $^1H$ spectrum of a N,N-dimethyl amide $((CH_3)_2NCOR)$ shows two sharp methyl resonances in a 1 : 1 ratio at 3.05 and 2.85 ppm. On warming the solution, the two resonances broaden and merge together at 84°C, above which only a single resonance is observed. Explain these observations. How would you expect the spectrum to appear if acquired at 500 MHz and 84°C? What is the rate constant and free energy of activation for the process measured at 84°C at 200 MHz?

## Further reading

1.   For a more detailed explanation of the physical basis of NMR see P. J. Hore, *Nuclear Magnetic Resonance*, Oxford Chemistry Primers, Oxford University Press, Oxford, 1995.
2.   R. Abraham, J. Fisher, and P. Loftus, *Introduction to NMR Spectroscopy*, Wiley, Chichester, 1988.
3.   H. Günter, *NMR Spectroscopy*, 2nd Ed., Wiley, Chichester, 1995.
4.   E. Breitmaier, *Structure Elucidation by NMR in Organic Chemistry. A Practical Guide.*, Wiley, Chichester, 1993.

# 5   Nuclear magnetic resonance spectroscopy: further topics

The previous chapter has introduced the essential material for applying ¹H NMR to problems of chemical structure. However, the power of modern pulse-Fourier transform NMR makes available to us a variety of techniques which assist us in the interpretation of spectra, in the subsequent determination of structures, and allows the routine observation of nuclei other than protons, as we now discover.

## 5.1   The pulse-Fourier transform approach

The idea of exciting a sample with an intense burst (or *pulse*) of electromagnetic radiation, followed by Fourier analysis of the emitted signals, has already been introduced in Chapter 1. This approach provides us with two major benefits over the continuous wave approach. The first, which was the original motivation for establishing the pulse-FT methodology, is one of time efficiency; many scans may be accumulated and added with a concomitant increase in signal-to-noise ratio for a given period of data collection. The second, which has been developed since the early 1980s, is the possibility of using more than one pulse prior to data acquisition, allowing one to control the behaviour of nuclear spins and hence the information presented in the resulting spectra. There now exists a spectacular array of these so-called *multi-pulse experiments*, a few of which we shall encounter in this chapter.

### The vector model of NMR

The previous chapter introduced how a single nucleus behaves in a static magnetic field. We now consider how an ensemble of nuclei behave, and the processes that occur following pulse excitation of a sample which ultimately produce the spectrum. In this pictorial approach, we shall use the vector model of NMR.

   Consider the magnetization arising from a collection of equivalent spin-$\frac{1}{2}$ nuclei. According to the Boltzmann distribution there will be a slight excess of spins in the $\alpha$ orientation, that is, parallel to the applied field, that will result in a *bulk magnetization vector*, *M*, along the $+z$ axis (which we define as being parallel to the applied field, Fig. 5.1) which behaves according to the rules of classical mechanics. Magnetization that exists on the $\pm z$ axis is referred to as *longitudinal magnetization*. We now wish to consider what happens when the sample is influenced by the application of a second magnetic field ($B_1$) associated with the radiofrequency (rf) radiation of the transmitter pulse. This is delivered through the coil, the geometry of which is such that the $B_1$ field is perpendicular to the static field, $B_0$. It becomes difficult to visualise the interaction of the oscillating rf with precessing spins so to help us in this we imagine the nuclear co-

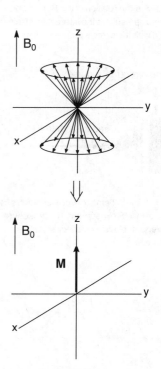

**Fig. 5.1**   An ensemble of spins gives rise to a *bulk magnetization vector*, *M*, which, at thermal equilibrium is parallel to the applied field.

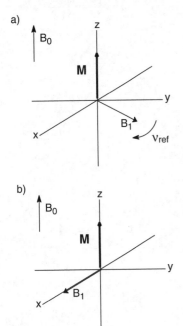

Fig. 5.2 Changing from the *laboratory frame* of reference (a) to the *rotating frame* of reference (b) 'freezes' the rotation of the $B_1$ field and helps us visualize the interaction of this field with the sample magnetization.

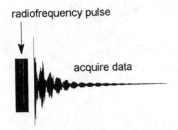

Fig. 5.3 Schematic representation of a simple NMR experiment. The procedure is repeated many times for signal averaging.

ordinate system to be rotating at the same frequency as the rf. This has the effect of 'freezing' the rf oscillations so the new $B_1$ field is now static in this co-ordinate system. Technically, this is referred to as changing from the *laboratory frame* of reference to the *rotating frame* of reference (Fig. 5.2) and a simple analogy would be you watching a child going around on a merry-go-round, which equates to observing events in the laboratory frame of reference, and then stepping onto the merry-go-round. In the latter case, the motion of the child is no longer apparent to you as he does not move toward you or away from you; you are now observing events in the rotating frame of reference which has simplified your *perception* of the motion. Furthermore, in the rotating frame we need only concern ourselves with frequency offsets from the transmitter frequency.

We are now in a position to consider the effect of the pulse in a simple pulse-acquire experiment (Fig. 5.3). Under the influence of the $B_1$ field the longitudinal magnetization is rotated toward the $x$–$y$ plane, by an amount dependent on the strength of the field and the duration of the pulse, to produce *transverse magnetization* (Fig. 5.4). A pulse which places $M$ exactly in the $x$–$y$ plane is referred to as a **90°** (or $\frac{\pi}{2}$) pulse and this corresponds to *equalizing* the populations of the $\alpha$ and $\beta$ states, whereas a pulse that inverts $M$ is a **180°** (or $\pi$) pulse, corresponding to *inversion* of the $\alpha$ and $\beta$ populations.

Any magnetization that is in the $x$–$y$ plane will in reality, that is, in the laboratory frame, be rotating at its Larmor frequency, and so will induce an oscillating voltage in the coil (much like the rotating magnet in a bicycle dynamo produces a voltage in the surrounding coil), which is the signal we wish to detect. Only transverse magnetization is able to induce a voltage in the coil, so that the maximum signal is achieved with a 90° pulse but no signal is observed for a 180° pulse.

Now, magnetization will not continue in the $x$–$y$ plane, or along the $-z$ axis, indefinitely, as this corresponds to deviation from the equilibrium populations, but will gradually return to the $+z$-axis (Fig. 5.5a), by losing its excess energy, that is, by *relaxing*. Thus, the oscillating voltage will decay away with time, producing the *free induction decay* (FID) (Fig. 5.5b) that may be Fourier transformed to produce the frequency spectrum. Typically, the FID is made up of a superposition of many oscillating signals arising from nuclei with different chemical shifts and coupling constants (see Section 1.3 also).

## 5.2   Relaxation

### Longitudinal relaxation

Longitudinal relaxation is the process which restores magnetization to the $+z$ axis (Fig. 5.6), or, in other words, restores the equilibrium population differences between the high and low energy spin states. The fundamental requirement for the relaxation of perturbed spins is the presence of an oscillating (time dependent) magnetic field at the appropriate frequency (the Larmor frequency), as is required for excitation. Whilst there are a number of mechanisms that can produce such a field, the most important is the *dipole–dipole interaction* which is the direct magnetic (through-space) interaction between two spins (Section 4.4). As a molecule tumbles in

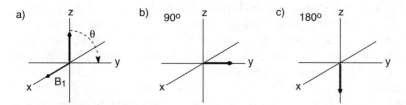

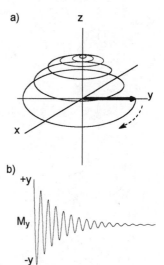

**Fig. 5.4** (a) Magnetization is rotated through an angle $\theta$ which is dependent on the magnitude of $B_1$ and the duration of the pulse, (b) $\theta = 90°$, (c) $\theta = 180°$.

solution, the relative orientation of the two dipolar-coupled spins changes, so that the field experienced at one nucleus due to its neighbour fluctuates (Fig. 5.7). This fluctuating field then provides a suitable means of relaxation for the spin. The energy lost by the spins is given out in the form of heat, although because the energies involved are so small, temperature changes are undetectable. Because the energy is distributed into the surroundings, known as the *lattice*, this form of relaxation is also referred to as *spin-lattice relaxation*.

For spin-$\frac{1}{2}$ nuclei, the return of $z$-magnetization follows a first-order (exponential) recovery characterized by the time constant $T_1$. Therefore, transverse magnetization produced by a 90° pulse requires a period of approximately $5T_1$ to relax back to its equilibrium state (after this period the magnetization has recovered by 99.93%). This has fundamental implications for the acquisition of pulse NMR spectra. If we apply pulses for successive scans too rapidly, the magnetization does not have time to recover, and ultimately we are left with no population difference to sample, so that further pulse-acquisitions will provide no more signal. This situation is known as *saturation* and it clearly defeats the point of signal averaging, so must be avoided. Proton $T_1$s for small organic molecules typically fall in the 0.5–5 s range, so acquisitions with a series of 90° pulses will usually require inter-pulse delays of 20 or so seconds. This is not the most time efficient way to collect data, so in practice spectra are routinely acquired with an excitation of only 30–60°, thus reducing the time needed for relaxation and allowing more acquisitions to be collected in a given time.

**Fig. 5.5** Magnetization will precess in the x–y plane, but also return to the +z axis through relaxation (a) and this process will induce a decaying oscillating voltage in the coil that is the *free induction decay* (b).

### Transverse relaxation

Transverse relaxation corresponds to the loss of bulk magnetization in the $x$–$y$ plane, as represented in Fig. 5.8. It does not relate to the restoration

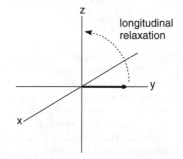

**Fig. 5.6** Longitudinal relaxation restores equilibrium populations and hence +z magnetization.

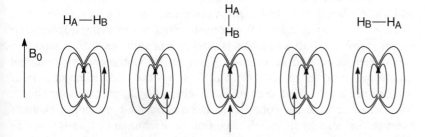

**Fig. 5.7** The field experienced at one nucleus due to its dipolar interaction with another fluctuates as the molecule tumbles in solution. This provides a mechanism for relaxation.

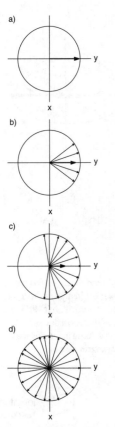

a)

b)

c)

d)

**Fig. 5.8** Transverse relaxation. The initial bulk magnetization vector (a) decreases as the many spins giving rise to it dephase (b-d) due to differences in local magnetic fields throughout the sample. Finally there is a random distribution of spins about the $x - y$ plane (d) and no net transverse magnetization.

$T_2^*$ includes a term for genuine transverse relaxation and broadening from magnetic field inhomogeneity.

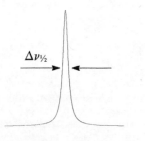

$\Delta\nu_{1/2}$

**Fig. 5.9** Linewidth is defined as the width of the line, in hertz, at half the resonance height.

of equilibrium populations so is distinct from longitudinal relaxation.

Following a 90° pulse, bulk magnetization exists in the $x$–$y$ plane as a single vector. Remember this is a macroscopic representation of very many aligned spins (they are said to have ***phase coherence*** at this point) which, for those with the same chemical shift, should precess together at the same frequency. In reality, some of these spins will experience a slightly reduced field whereas others experience a slightly greater field than the average, brought about by small variations in the local magnetic environments. This leads to small changes in their precessional frequencies, so that some fall behind the bulk vector whereas others creep ahead of it (Fig. 5.8b–d). This then leads to a 'fanning-out' of the magnetization vector as the phase coherence is lost, until ultimately the spins are evenly distributed about the $x$–$y$ plane and there is no resultant magnetization and hence no detectable signal. The distribution of precessional frequencies corresponds to a spread of frequencies in the final spectrum, so that individual resonances are not infinitely narrow lines, but have a defined linewidth (Fig. 5.9). Faster blurring of the magnetization following excitation implies greater frequency differences between individual spins, and hence broader lines in the final spectrum. This relaxation is again exponential for spin-$\frac{1}{2}$ nuclei and characterized by the time constant $T_2$, which, as expected from the above arguments, is inversely proportional to the linewidth.

Mechanisms for this relaxation are essentially similar to those for longitudinal relaxation, but are beyond the scope of this book. Transverse relaxation does not involve a loss of energy to the surroundings but energy is retained by the spins and transferred amongst themselves by mutual spin-flipping (such as $\alpha\beta \leftrightarrow \beta\alpha$) so is also referred to as ***spin-spin relaxation***. For small molecules, $T_2 = T_1$ and in any circumstance $T_2$ can never be greater than $T_1$ as clearly there can be no magnetization left in the $x$–$y$ plane if it has all returned to the $+z$ axis.

The dominant cause of blurring of the magnetization vector, and hence line-broadening in NMR spectra, actually arises not from genuine transverse relaxation but from non-uniformity in the magnetic field, meaning spins in different regions of the sample experience slightly different fields. The ***linewidth*** of a NMR resonance ($\Delta\nu_{1/2}$, Fig. 5.9) is then given by

$$\Delta\nu_{1/2} = \frac{1}{\pi T_2^*}$$

Two mechanisms for relaxation are also considered briefly now, which both result in broadening of resonances. The first is ***quadrupolar relaxation*** which arises because the spins of nuclei with $I > \frac{1}{2}$ (quadrupolar nuclei) are efficiently relaxed by ***electric field gradients*** about the nucleus. Whilst, in organic chemistry, we are not often interested in observing quadrupolar nuclei themselves (although this is possible) protons that are coupled to them often appear to be unusually broadened. Commonly encountered quadrupolar nuclei include $^2$H and $^{14}$N, so that, for example, amide protons are often broad, even in the absence of dynamic exchange effects. This arises because the coupling to the quadrupolar nucleus is *partially* averaged to zero by its rapid relaxation i.e. exchange of spin-states; this may be considered to be similar to the exchange decoupling of acidic protons discussed in Section 4.6. The second mechanism is relaxation by unpaired electrons, such as in paramagnetic species. The magnetic moment of the electron is approximately $10^3$ times that of the proton, so it produces a strong fluctuating field and efficient relaxation. The rapid relaxation in turn causes *paramagnetic broadening* of NMR resonances.

## 5.3  The nuclear Overhauser effect

The nuclear Overhauser effect (NOE) holds a position of great importance in organic structure elucidation as it enables us to define the three-dimensional stereochemistry of our molecules of interest. The NOE is the change in signal intensity of a resonance when the population differences of a near-neighbour are perturbed from their thermal equilibrium values, such as by the application of a radiofrequency field, as described below. The effect is transmitted *through space via* the direct magnetic influence of one nucleus on another (the dipolar interaction) and *does not* require or depend on the presence of scalar (through-bond) coupling. Its usefulness in defining molecular geometry arises because the NOE has a dependence on, amongst other things, the distance between interacting nuclei. For these effects to be observable the internuclear distance must be less than *ca.* 0.4 nm so that the presence of the NOE places boundary limits on the distance between nuclei and a collection of such NOEs within a molecule can often be sufficient to define its stereochemistry. Thus, a combination of information arising from through bond (scalar coupling) and through space (NOE) interactions may be combined to define the structure and stereochemistry of a molecule.

### Origin of the NOE

Consider two spins that are close in space and hence share a mutual dipolar coupling, giving rise to the energy level diagram of Fig. 5.10a. In many text books the spins are generally referred to as I and S for historical reasons, so we shall also adopt this terminology. The total spin population in the system is $4N$. The lowest energy level contains an excess of spins, $\Delta$, the highest energy level contains a deficit of $\Delta$ and the middle levels have, to a good approximation, equal populations, giving rise to a population difference across all transitions of $\Delta$. Now suppose that both the S transitions are saturated, that is, the population differences are forced to zero (Fig. 5.10b), such as by applying a weak radiofrequency field at the S spin Larmor frequency. The S spin population differences are now no longer those of the original Boltzmann distribution so that, just like any chemical state that is forced away from thermal equilibrium, the system will alter itself to counteract the changes imposed. The population differences across the I transitions have not been altered by the saturation of S (Fig. 5.10b) so transitions involving a single I spin, such as $\alpha\alpha \leftrightarrow \beta\alpha$ ($W_1$ transitions, where the subscript indicates the overall change in the magnetic quantum number for the transition) will play no part in re-establishing equilibrium. Instead the processes indicated by $W_0$ and $W_2$ must be operative. These transitions are not directly observable in NMR spectra, that is they do not produce a signal in the NMR coil, as they do not obey the selection rule $\Delta M = \pm 1$, where $M$ is the magnetic quantum number. However, they can operate to change spin states, as is required in

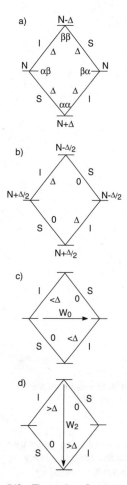

**Fig. 5.10**  The nuclear Overhauser effect. Energy levels for dipolar coupling between two spins, I and S. Inside each transition is shown the population difference across it. (a) At equilibrium the $\alpha\alpha$ state has an excess, whereas the $\beta\beta$ state has a deficit, of spins giving rise to a population difference of $\Delta$ for all transitions. (b) Immediately after saturating the S spins. (c and d) The systems attempts to re-establish population differences across the S transitions by relaxation *via* the $W_0$ (c) and $W_2$ pathways (d). The new I populations that result produce I spin signal intensities that are different from their equilibrium values. Throughout, the S spin population differences are held at zero (see text also).

this case *via* dipole-dipole relaxation. If the $W_0$ process is operative (Fig. 5.10c) the spin populations are depleted at the bottom of one I transition and enhanced at the top of the other, so that the I spin population *differences* are *decreased*, leading to a net reduction in signal intensity if we record the spectrum at this point. This is referred to as a ***negative NOE***. If the $W_2$ process occurs, (Fig. 5.10d) then one I transition is depleted at the top whilst the other is enhanced at the bottom, and the I transition population *differences* are *increased*. This enhances the I signal intensity and this is then a ***positive NOE***. Thus, by saturating the transitions of the S spins, we have altered the signal intensity of the I spins.

For the changes in signal intensity to occur, spin states must alter and this occurs *via* the dipole-dipole relaxation mechanism. Thus, the NOE is intimately related to longitudinal relaxation and hence molecular motion and the sign of an observed NOE is a balance between the $W_0$ and $W_2$ pathways. $W_0$ represents a low energy transition (there is a small energy difference between the $\alpha\beta$ and $\beta\alpha$ states) so that low frequency molecular tumbling can provide a suitable oscillating field for relaxation and large molecules, such as proteins, tend to show negative NOEs. $W_2$ on the other hand is a high energy process that is strongly influenced by high frequency tumbling and most molecules of interest to organic chemists give rise to positive NOEs.

The rates at which the above population changes occur depend on $r^{-6}$, where $r$ is the internuclear distance, and it is through this that the NOE itself has a distance dependence; the closer the nuclei, the faster the effect will build up and the stronger it is likely to be when we come to sample the new populations. The power of –6 ensures the effect falls rapidly with distance.

The NOE enhancements, $\eta$, are defined as

$$\eta = \frac{I - I_0}{I_0} \times 100\%$$

*I* is the resonance intensity in the presence of the NOE.
$I_0$ is the resonance intensity in the absence of the NOE.

$\gamma_S$ represents the magnetogyric ratio of the saturated spin and $\gamma_I$ that of the observed spin

The maximum magnitude of the NOE ($\eta_{max}$) between two spin-$\frac{1}{2}$ nuclei in a small, rapidly tumbling molecule is given by the simple equation

$$\eta_{max} = \frac{1}{2}\frac{\gamma_S}{\gamma_I} \times 100\%$$

For proton-proton NOEs this value is therefore 50%, whereas enhancements of hundreds of percent may be obtained when irradiating protons and observing a lower frequency nucleus, such as $^{13}C$ (Section 5.4). In routine $^1H$–$^1H$ NOE experiments enhancements of around 1–20% are most often observed. A simple example of the differentiation of $E$ and $Z$ alkenes by comparing their NOE data may be seen in Fig. 5.11.

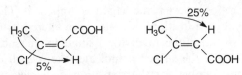

**Fig. 5.11**   The NOE can be used to distinguish between $E$ and $Z$ alkenes.

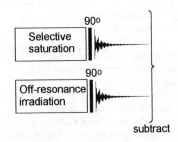

**Fig. 5.12**   A schematic representation of the NOE difference experiment.

### The NOE difference experiment

In this experiment (Fig. 5.12), a radiofrequency field is applied at the frequency of the S spins saturating them for a period that is many times

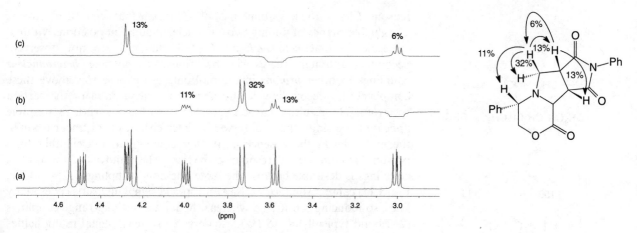

**Fig. 5.13** NOE difference experiments for the molecule shown (selected data only). The partial $^1$H spectrum (a), and the results following saturation (b) and (c). Negative-going signals indicate the position of saturation.

the $T_1$ of the I spins, so that there is sufficient time for the new population differences to become established. The proton spectrum is then acquired and the data stored. A second experiment represents the control where the applied frequency is away from all resonances, say at the very edge of the spectrum, so that no NOEs are generated. The two spectra are then subtracted so that everything cancels, except where signal enhancements have been generated (Fig. 5.13). The reason for the subtraction lies in the fact that we are looking for very small changes in signal intensities, often only a few percent, and these are better revealed in the difference spectrum. The resulting enhancements are termed *steady-state NOEs* because they were created whilst the system was held away from the equilibrium state by the application of the saturating radiofrequency.

## 5.4 $^{13}$C NMR spectroscopy

Despite the low natural abundance (only 1.1%) and relatively low sensitivity of the $^{13}$C nucleus, its observation is now made routine by the use of efficient signal averaging, made possible by the pulse-FT approach, and by the availability of higher magnetic field strengths. However, sample quantity requirements are still approximately ten times those for proton observation.

The appearance of the conventional carbon spectrum (Fig 5.14a) is markedly different from that of the proton spectrum for a number of

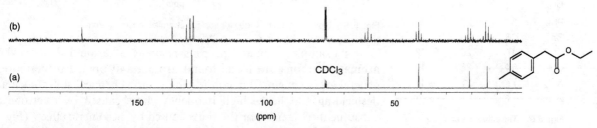

**Fig. 5.14** (a) the conventional broadband proton decoupled spectrum of ethyl *p*-tolyacetate, and (b) the full proton coupled spectrum acquired with the same number of scans. Notice the poorer signal-to-noise ratio in (b) (see text).

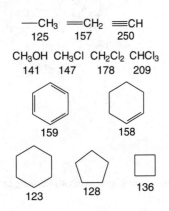

$-CH_3$    $=CH_2$    $\equiv CH$
  125      157      250

$CH_3OH$   $CH_3Cl$   $CH_2Cl_2$   $CHCl_3$
 141      147      178      209

159          158

123     128     136

**Fig. 5.15** Some examples of 1-bond H–C coupling constants (Hz). The dominant factor in determining the magnitude of these is the degree of 's-character' ($f_s$) of the carbon, and the general equation $^1J_{CH} \approx 500.f_s$ applies ($f_s$ is 0.25, 0.33 and 0.5 for sp$^3$, sp$^2$ and sp respectively). In addition, electronegative substituents tend to increase the magnitude of $J$.

reasons. Firstly, the low abundance of $^{13}$C means that there is, effectively, a negligible chance of finding two $^{13}$C nuclei in adjacent positions within a molecule so that **homonuclear** $^{13}$C–$^{13}$C couplings are not observed. Secondly, although there will be numerous possible **heteronuclear** couplings to other *protons* in the molecule, we choose to remove these completely by decoupling *all* protons during acquisition of the carbon spectrum. The resulting carbon spectrum then has a surprisingly simple appearance; a single line is observed for each chemically distinct carbon in the molecule, in the absence of further coupling to nuclei other than protons or coincidental resonance overlap. The removal of the proton couplings is desirable because the large one-bond couplings (125–250 Hz, Fig. 5.15) cause resonance overlap and spread signal intensities over many lines, so reducing sensitivity, whereas the numerous long-range couplings (2–3 bond typically < 15 Hz) also serve to increase signal multiplicities and further reduce sensitivity (Fig. 5.14b). All these detrimental factors are removed by the **broadband decoupling** of all protons, which is similar to the decoupling discussed in Section 4.4 except that the technique is modified so as to make it non-selective. In addition, a further gain in sensitivity may be obtained from the NOE that is generated on the carbons from the saturated protons, which may serve to *enhance* signals by as much as 200% ($0.5\gamma_H/\gamma_C$, Section 5.3 and Table 4.1). The 'singlet' appearance of carbon resonances and the large chemical shift range exhibited by carbon nuclei (see below) means that resonance overlap is rarely a problem and the carbon spectrum often provides a means of counting carbon nuclei within a molecule.

## Chemical shifts and resonance intensities

As mentioned above, the chemical shift range for the carbon-13 nucleus is much greater than that of the proton, covering a typical range of 0–220 ppm, again with the shift of tetramethylsilane assigned as 0.0 ppm (Fig. 5.16). A useful 'rule-of-thumb' for estimating chemical shifts is to multiply the shift of a proton in an equivalent chemical environment by 20.

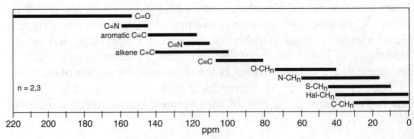

**Fig. 5.16** Typical $^{13}$C shift ranges for various chemical environments.

| X | $XC^1H_2C^2H_2C^3H_3$ | | |
|---|---|---|---|
| | $C^1H_2$ | $C^2H_2$ | $C^3H_3$ |
| Et | 34 | 22 | 14 |
| HOOC | 36 | 18 | 14 |
| SH | 27 | 27 | 13 |
| NH$_2$ | 44 | 27 | 11 |
| Phe | 39 | 25 | 15 |
| Br | 36 | 26 | 13 |
| Cl | 47 | 26 | 12 |
| OH | 64 | 26 | 10 |
| NO$_2$ | 77 | 21 | 11 |

**Fig. 5.17** The influence of electronegative substituents on $^{13}$C chemical shifts (see also Fig. 4.9).

For example, aromatic protons resonate at around 7 ppm whereas aromatic carbons are to be found around 140 ppm. Furthermore, the general arguments used in considering proton chemical shifts, such as the deshielding and hence high frequency shifts caused by electronegative substituents (Fig 5.17) or the shifts caused by mesomeric effects (Fig. 5.18 and see Fig. 4.10) can also be applied to carbon chemical shifts. Hybridization also plays an important role in carbon-13 NMR, and spectra can be conveniently divided into regions corresponding to sp$^3$, sp$^2$ and sp types although exceptions to this do occur. Sp$^3$ carbons tend to resonate at 0–50 ppm in the absence of electronegative substituents, but may be as high as 90 ppm with such substitutions (Fig. 5.19). The sp$^2$ carbons in alkenes and aromatics resonate in the 100–160 ppm region, whereas those of carbonyls are found, characteristically, at higher frequencies in the 160–220 ppm region. It is possible to further differentiate between types of carbonyl carbons; ketones and aldehydes generally occur above 180 ppm whereas acids, esters and amides fall below this. In all cases, conjugation of the C=O bond reduces the partial positive charge on the carbon and causes shifts to lower frequency. The sp carbons of alkynes resonate around 80–100 ppm, but are less frequently encountered. Extensive tables are available which predict carbon shifts for various substitution patterns (see Appendices also), and computer programs are available that essentially automate this process.

One notable effect in carbon-13 NMR is the so-called γ-effect in which a resonance moves to lower frequency when substitution is made at the γ position relative to it (Fig. 5.20). This arises from van der Waals contacts with the substituent, and so has maximum effect for ring structures or alkenes where conformation is restricted, and may be useful in stereochemical analysis.

Resonance intensities in *routine* carbon spectra are not as readily interpreted as those in proton NMR, and it is not usual to consider integrating spectra for even semi-quantitative analysis. This is due to three factors. The first is the unknown enhancement caused by the NOE which is unlikely to be the same for all resonances, the second is the long relaxation times of carbon nuclei ($T_1$s are often tens of seconds) so that populations are partially saturated to an unknown degree when the data are acquired and the third relates to the way in which the data is digitized, but lies beyond the scope of these discussions. Nevertheless, it is often possible to identify quaternary carbons as they usually have lower intensities than those bearing protons, largely due to the longer relaxation times of the non-protonated carbon.

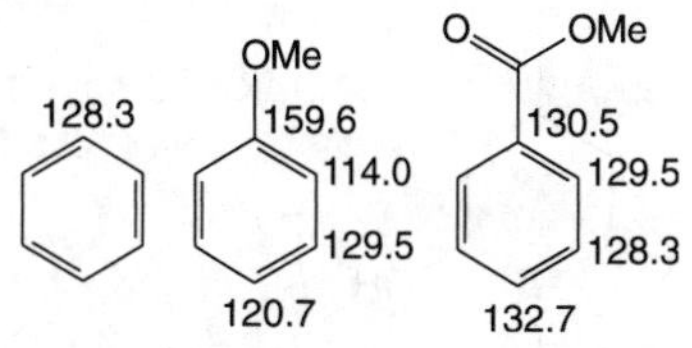

## Determining multiplicities

Despite the many advantages stated above for the broadband decoupling of protons, there is one major drawback,—all potentially useful information relating to heteronuclear coupling is lost. In particular, we are unable to determine whether the carbon resonance in question is that of a methyl, methylene, methine or quaternary functionality, which, in principle, could readily be obtained in a proton-coupled spectrum from the multiplicity arising from the large one-bond couplings, that is, quartet, triplet, doublet and singlet respectively (although further complicated by the significantly smaller long-range couplings). The traditional way to regain such information is termed **off-resonance decoupling** in which the proton decoupling frequency is placed to one end of the proton spectrum during data acquisition. This has the effect of removing the long-range couplings but *partially* retaining the one-bond couplings, which are scaled down by an unknown factor yet still retain their distinctive multiplicities. Clearly this still suffers from many of the disadvantages of the fully proton-coupled spectrum.

**Fig. 5.18** The influence of mesomeric effects on $^{13}$C chemical shifts (see also Fig. 4.10).

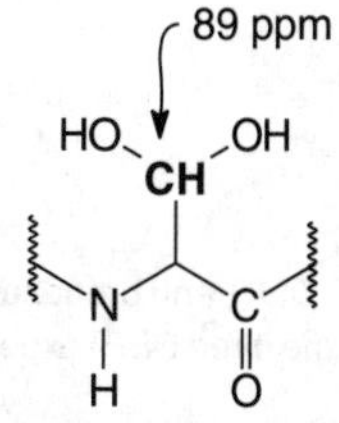

**Fig. 5.19** Sp$^3$ carbons can resonate close to the 'sp$^2$' region if they carry a number of electronegative substituents.

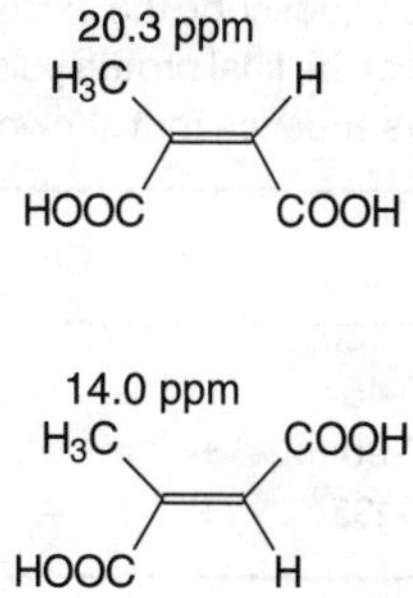

**Fig. 5.20** The γ-effect causes low frequency shifts for $^{13}$C resonances.

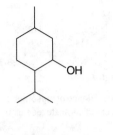

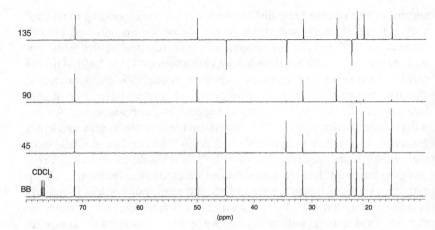

**Fig. 5.21**  DEPT and broadband decoupled spectra of menthol. Carbon multipliciites can be readily determined by inspection of the signal phases in the three DEPT experiments. A single DEPT-135 experiment is often sufficient in the absence of methyl groups.

**Table 5.1**  *Relative* signal phases observed in DEPT experiments. Carbons that carry no protons give no response. The numbers 45°, 90° and 135° describe the excitation angle of the final proton pulse in the pulse sequence (not shown).

|          | CH | CH$_2$ | CH$_3$ |
|----------|----|--------|--------|
| DEPT-45  | +  | +      | +      |
| DEPT-90  | +  | 0      | 0      |
| DEPT-135 | +  | −      | +      |

The modern way to determine carbon multiplicities is *via spectrum editing* methods. In these techniques, a multi-pulse sequence manipulates the spin systems so that spectra can be acquired with broadband decoupling but in which the resonances display signal phases, that is, positive or negative signal intensities, which are dependent on the number of directly attached protons. One technique now used routinely in organic chemistry is the *DEPT* (Distortionless Enhancement by Polarization Transfer) experiment for which the signal phases are presented in Table 5.1. An analysis of the three DEPT spectra, along with the broadband decoupled spectrum, is usually sufficient to determine all carbon multiplicities (Fig. 5.21), and is invaluable in the routine interpretation of carbon spectra. Alternatively the spectra may be combined by computer manipulation to present sub-spectra for CH, CH$_2$ and CH$_3$ resonances only, hence the term spectrum editing.

## 5.5  Two-dimensional NMR spectroscopy

The spectra that have been considered so far are all termed one-dimensional spectra because all information (chemical shifts and scalar couplings) is displayed along only one frequency axis. We now wish to consider a class of experiments, two-dimensional (2D) spectroscopy, in which information exists along a second frequency axis in addition to the first. Responses in the 2D plot enable us to correlate resonances on the first axis with those on the second, and this provides us with information on how spins are related to one another within a molecule. Generally, correlations in the 2D plot indicate the presence of scalar coupling, dipolar coupling (an NOE) or chemical exchange between spins, depending on the exact nature of the experiment. The first of these has most importance in organic chemistry, and shall be considered here, in two slightly different guises, but first we consider how a two-dimensional spectrum arises.

### Introduction to the second dimension

All two-dimensional experiments follow the scheme of Fig. 5.22. The preparation, P, and mixing, M, segments are pulses, or clusters of pulses of some description, whilst the evolution, E, and detection, D, segments are time periods in which the magnetization simply precesses. During the detection period, the free induction decay (FID) is collected to produce one time-domain data set, as in the one-dimensional experiment (Section 5.1), which may be Fourier transformed to generate one frequency axis. The evolution period is incremented sequentially throughout the experiment and thus creates a second time-domain, which ultimately leads to the second frequency axis. To see how this is achieved, we consider the simple case for which P and M are both 90° pulses (indeed, this was the first two-dimensional experiment proposed, Fig. 5.23 and see below) and we shall assume our sample contains only identical, uncoupled spins.

Following the fate of our single magnetization vector (Fig 5.24) we see the first 90° pulse places the magnetization in the $x$–$y$ plane, after which it precesses for a time period, $t_1$ (not to be confused with $T_1$ for the longitudinal relaxation time, Section 5.2) according to its characteristic Larmor frequency. The second 90° pulse rotates any component of the magnetization remaining along the y-axis onto the $-z$-axis, but leaves the $x$-magnetization component unaffected as the $B_1$ field along $x$ has no influence on $x$-magnetization. The remaining $x$-magnetization is then detected as it continues to precess, and the resulting FID stored, whereas the $z$-magnetization is unobservable (Section 5.1).

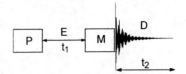

**Fig. 5.22** Schematic illustration of all two-dimensional NMR experiments. Four periods can be identified; P, preparation, E, evolution, M, mixing and D, detection.

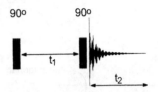

**Fig. 5.23** A simple 2D NMR experiment in which preparation and mixing are carried out by single 90° pulses.

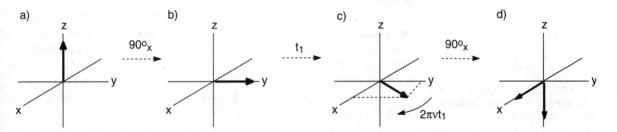

**Fig. 5.24** The fate of a single magnetization vector during our simple 2D pulse sequence. Following excitation, the vector evolves at a frequency, $\nu$, and through an angle $2\pi\nu t_1$ radians prior to the second pulse. Only magnetization in the x–y plane is detected.

To generate another time-domain, the experiment is simply repeated with increasing values of $t_1$, starting from $t_1 = 0$, and the FID for each experiment stored separately (Fig. 5.25). With zero evolution time, the two sequential 90° pulses place the initial $+z$-magnetization directly onto the $-z$ axis, so that the first FID contains no signal. As $t_1$ becomes significant, $x$-magnetization has time to develop, so that later FIDs will contain signals. If each of the collected FIDs were transformed, the result would be that of Fig. 5.26a. As stated, the first spectrum would contain no signal, maximum peak intensity is obtained when the magnetization vector has evolved through 90° (all magnetization on $+x$ axis, none on $z$ following the second pulse), passes through a null when the vector has evolved through 180° (all magnetization on the $+z$ axis following the second pulse) and so on. In later FIDs, the intensity of the detected signal will decrease as relaxation during the $t_1$ period causes the magnetization vector to decay. Thus, if we follow the intensity of the resonance with respect to the $t_1$ time period, we see it simply represents a decaying

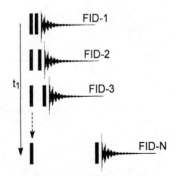

**Fig. 5.25** The generation of a 2D data set. The experiment is repeated with the $t_1$ time period incremented each time and the FIDs stored.

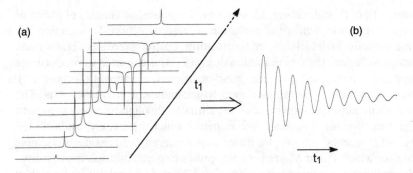

**(a)**

**(b)**

$t_1$

$t_1$

**Fig. 5.26**   Fourier transformation of the FIDs collected during the 2D experiment yields a series of 1D spectra (a). Following the intensity of this resonance, we see it varies sinusoidally and decays with time (b).

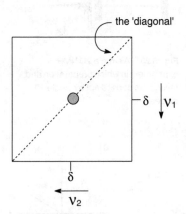

the 'diagonal'

$\delta \quad \big\downarrow \nu_1$

$\delta$

$\nu_2$

**Fig. 5.27**   Schematic representation of a 2D spectrum for a sample containing identical, uncoupled protons.

oscillating signal (Fig. 5.26b); *it is a free induction decay for the $t_1$ time period*. The frequency of the oscillation is the frequency of precession, $\nu$, during the evolution period; we say the detected magnetization has been *frequency labelled*.

We may now Fourier transform the FIDs with respect to $t_1$ to obtain another frequency dimension, so producing a two-dimensional frequency plot (Fig. 5.27). The basic procedure of incrementing a time period and repeating the experiment is an operation fundamental to all 2D NMR experiments.

The schematic spectrum of Fig 5.27 in itself provides us with no information we could not otherwise obtain from the one-dimensional spectrum, as the information in both dimensions is identical, simply the chemical shift of the single resonance. Two-dimensional NMR becomes useful in the presence of different spins when the information represented in one dimension correlates with *different* information in the other, as we shall see below.

## Proton-proton correlation

As has been discussed in the previous chapter, the identification of coupling between nuclei is crucial in defining a molecular structure. Two-dimensional proton-proton chemical shift correlation spectroscopy (*COSY*) provides a map of all coupling networks between protons in a molecule in a single experiment. Because we are correlating like spins, this is one example of *homonuclear correlation spectroscopy*. The COSY sequence is actually that described above (Fig. 5.23), and, although it was the first 2D experiment proposed (in 1971) it is still the most widely used. Now let us consider, briefly, the result of performing the COSY experiment on two groups of spins, A and X, that are coupled to each other. Taking the frequency labelling part described above to be implicit, the additional feature in this instance is the action of the second 90° pulse on the coupled spins. This is able to transfer magnetization partially from one spin to its *J*-coupled partner, so that magnetization which evolved as A magnetization at frequency $\nu_A$ in $t_1$, then evolves as X magnetization at frequency $\nu_X$ during $t_2$. This process is known formally as *coherence transfer* and is only possible because of the presence of scalar coupling between spins. Similarly, the reverse process can occur, that is magnetization transfer from X to A, and in addition some part of the original magnetization will not be transferred by the pulse so will precess at the same frequency in both time periods. Thus, two types of response may be identified in the two-dimensional plot (Fig. 5.28). The first, known as *diagonal peaks*, represents magnetization that has evolved at the same frequency in both time periods, and provides no new information (the hypothetical line corresponding to $\nu_1 = \nu_2$ is referred to as the *diagonal*). The second, known as *cross peaks* (or off-diagonal peaks), represents magnetization that has evolved at two different frequencies in the two time periods. These provide the results of interest, as their presence is direct evidence for coupling between the protons that resonate at the two correlated chemical shifts. Because magnetization may be transferred either way (A→X and X→A) cross peaks are symmetrical about the diagonal.

The use of the COSY experiment allows us to rapidly determine the coupling pathways within even relatively complex molecules, by correlating the chemicals shifts of coupled spins in a stepwise manner. An example and its interpretation is presented in Fig. 5.29. Similar information can, in

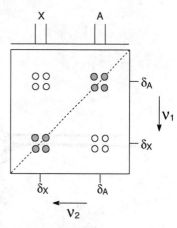

**Fig. 5.28** Schematic representation of a 2D spectrum for a coupled, two-spin AX system. Filled circles represent diagonnal-peaks ($\nu_1 = \nu_2$), open circles, cross-peaks ($\nu_1 \neq \nu_2$). Above is a schematic representation of the 1D spectrum illustrating the doublet fine-structure of each resonance.

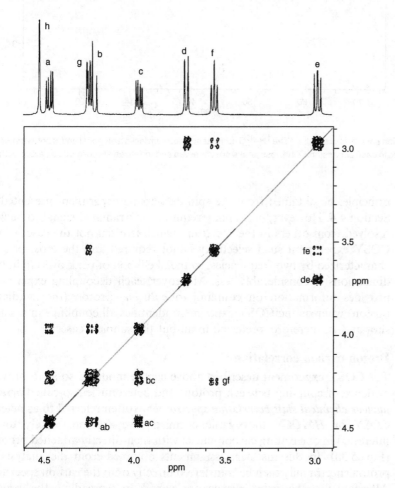

**Fig. 5.29** 2D proton-proton shift correlation spectrum (COSY) of the molecule shown (excluding the phenyl groups). The spectrum is conventionally presented as a contour plot in which correlations between protons can be traced in a step-wise manner. The fine-structure within the crosspeaks is due to partially resolved spin-spin couplings.

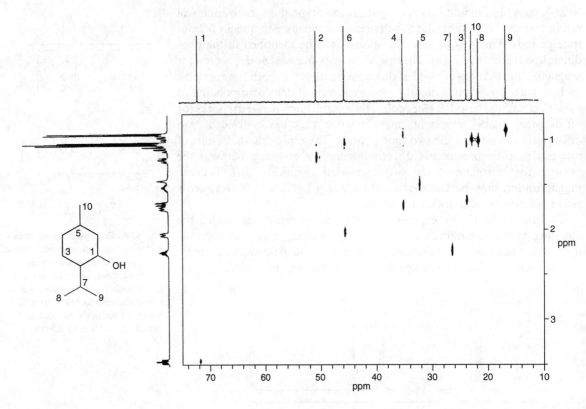

**Fig. 5.30**   2D proton-carbon shift correlation spectrum of menthol. The $^1$H–$^{13}$C correlation spectrum lacks a diagonal and is not symmetrical because different frequencies are displayed in each dimension. Only cross peaks are observed and these readily correlate a carbon with its directly bonded proton(s).

principle, be obtained from the spin decoupling experiment presented in Section 4.4. However, decoupling requires the irradiated signal to be well resolved from others in the spectrum, which is often not the case. In the COSY experiment such selectivity is not required and the cross-peak is characterized by two frequencies, so the likelihood of peak overlap in two dimensions is considerably less. Moreover, each decoupling experiment provides information on coupling to *only one* proton (the irradiated proton) whereas the COSY experiment identifies all couplings in a single shot and is, therefore, preferred in all but the simplest cases.

**Proton-carbon correlation**

The COSY experiment described above can be modified so as to provide evidence of coupling between protons and heteronuclei, so-called ***heteronuclear chemical shift correlation spectroscopy*** (often referred to as hetero-COSY or ***HMQC***). In organic chemistry we are most likely to be interested in correlating carbon nuclei with their directly attached protons (Fig. 5.30). In this manner, assignments obtained from the analysis of proton spectra may then be transferred directly onto the carbon spectrum. Alternatively, the extra dispersion gained by spreading the proton spectrum along the $^{13}$C dimension of the two-dimensional spectrum, may assist in the interpretation of the proton spectrum itself.

## 5.6  Exercises

1.  Assign the $^{13}C$ spectrum of ethyl *p*-tolylacetate using the spectra shown in Fig. 5.14.

2.  An unknown compound has a molecular formular of $C_7H_{10}O$ and gives responses in its $^{13}C$ spectra as summarized below. Use these data to assign the resonances as C, CH, $CH_2$ and $CH_3$ groups and, given that the molecule absorbs strongly in the infra-red region at 1670 $cm^{-1}$, suggest a possible structure for this molecule.

|                        | 1   | 2   | 3   | 4   | 5   | 6   | 7   |
|------------------------|-----|-----|-----|-----|-----|-----|-----|
| $^{13}C$ Shift/ppm     | 24  | 26  | 32  | 38  | 127 | 162 | 205 |
| DEPT-45                | +   | +   | +   | +   | +   | 0   | 0   |
| DEPT-90                | 0   | 0   | 0   | 0   | +   | 0   | 0   |
| DEPT-135               | −   | +   | −   | −   | +   | 0   | 0   |

Chemical shifts were determined from the broadband decoupled carbon spectrum, and DEPT responses are given as: + = positive signal phase, − = negative signal phase and 0 = no signal.

3.  The table below summarizes the results of NOE difference experiments performed on the molecule shown. Using these results, and the knowledge that the C5 stereochemistry is as shown, determine the stereochemistry at the C2 and C3 positions.

| Saturated Proton | NOE enhancements (%) | | | |
|------------------|------|------|------|--------|
|                  | H2   | H3   | H5   | $CH_3$ |
| H2               | X    | 1.2  | 2.6  | 3.6    |
| H3               | 2.9  | X    | −    | 2.6    |
| H5               | 3.6  | −    | X    | −      |
| $CH_3$           | 9.4  | 7.9  | −    | X      |

## Further reading

1.  A. E. Derome, *Modern NMR Techniques for Chemistry Research*, Pergamon Press, Oxford, 1987.
2.  J. K. M. Sanders and B. K. Hunter, *Modern NMR Spectroscopy,—A Guide for Chemists*, 2nd Ed., Oxford University Press, Oxford, 1993.
3.  R. K. Harris, *Nuclear Magnetic Resonance Spectroscopy*, Longman, Harlow, 1986.
4.  D. Neuhaus and M. P. Williamson, *The Nuclear Overhauser Effect in Structural and Conformational Analysis*, VCH Publishers, Weinheim, 1989.
5.  H. Friebolin, *One and Two-Dimensional NMR Spectroscopy*, 2nd. Ed., VCH Publishers, Weinheim, 1993.

# 6   Mass spectrometry

## 6.1   Introduction

Having dealt with techniques which involve the absorption of electromagnetic radiation, we will now consider a totally different type of
analysis which enables the mass of the molecule to be determined and also
provides further information about how it is constructed.

In the basic mass spectrometric experiment (first demonstrated by Wien
in 1898), individual molecules of a sample are bombarded with high
energy electrons which cause the ejection of one or more electrons from
the substrate on impact. The technique is known as ***electron impact
ionization*** (Fig. 6.1).

high energy electron

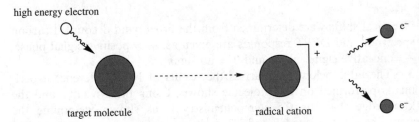

target molecule                                          radical cation

**Fig. 6.1**   Ionization of molecules by impact with high energy electrons.

The positively charged ions are then accelerated along an electrical
potential and pass through a magnetic field which causes them to be
deflected to an extent which is dependent upon the charge on the molecule
and the magnetic field strength and is inversely dependent upon the
momentum (and hence the mass) of the positively charged species (Fig.
6.2). The actual relationship is:

$$\frac{m}{z} = \frac{B^2 R^2}{2V}$$

$m$ = mass of the ion
$z$ = charge on the ion
$B$ = applied magnetic field
$R$ = radius of arc of deflection
$V$ = applied accelerating voltage.

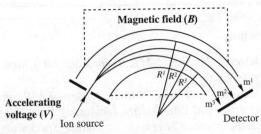

**Fig. 6.2**   Accelerated ions are deflected by a magnetic field.

Heavy ions are deflected less than light
ions under the same conditions.

For the experiment to work there must be no intermolecular exchanges of
energy and so extremely high vacua ($ca\ 10^{-6}$ torr) are required.
Measurement of $m/z$ permits the mass of each charged species to be
ascertained if the charge is known. In such an arrangement the highest

$m/z$ will correspond to the singly charged intact molecule (the ***molecular ion***) and gives the molecular mass directly. Sometimes however, peaks of non-integral mass are detected and these correspond to multiply charged species. In mass spectrometry, individual ionized molecules are measured in contrast to UV, IR, or NMR spectroscopy in which the bulk sample is analysed. Consequently it is possible to distinguish molecules containing different isotopes such as $^{35}Cl$ and $^{37}Cl$.

In addition to producing the intact radical cation, impact of an electron may cause sufficient energy to be transferred to the vibrational modes of the molecule that it undergoes bond rupture, giving rise to cleavage fragments which themselves may still possess sufficient vibrational energy to undergo subsequent fragmentation (Fig. 6.3). Those fragments which retain the positive charge will also be deflected in the magnetic field but to a greater extent than the intact ions due to their lower mass.

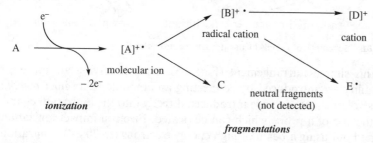

**Fig. 6.3** Ionization and possible fragmentation sequences.

The signal generated at the detector is dependent upon the number of ions of any particular mass arriving at the analyser. It follows therefore that the most intense signals in a mass spectrum will arise from the cleavages which occur most readily. Such cleavages are likely to occur at weak bonds, or to give stabilized fragments, and so inspection of the more intense fragmentation peaks can give us insight into the structure of the molecule. The mass spectrum of a compound is usually displayed as a bar chart of ion intensity against increasing integral mass, with the strongest peak (the ***base peak***) normalized to 100% intensity (Fig. 6.4).

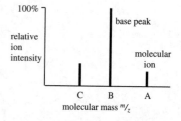

**Fig. 6.4** Representation of a mass spectrum.

One of the great advantages of mass spectrometry is that it requires extremely small amounts of material (a nanogram can be sufficient) and so can be considered effectively to be a non-destructive method of analysis.

The basic instrument which analyses the ions using a magnetic field is referred to as a ***single focusing mass spectrometer***, but a common alternative arrangement uses an ***electrostatic field***, either before or after passage through a magnetic analyser, when the spectrometer is referred to as ***double focusing***.

In addition to ***electron impact ionization***, a useful commonly encountered ionization technique is ***chemical ionization*** where the substrate is bombarded with positively charged atoms or molecules instead of electrons (Section 6.3). This milder technique means that molecules which are unable to survive the very rigorous conditions of electron impact intact may still yield information about their molecular weight. Commonly ammonia is used as the source of primary ions

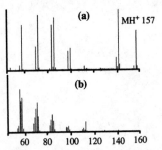

**Fig. 6.5** Mass spectra of cyclodecanol (MW = 156) (a) NH$_3$ chemical ionization (b) electron impact.

resulting in addition of H$^+$ and NH$_4^+$ to the substrate molecule to give *quasi–molecular ion* peaks at M + 1 (Fig. 6.5) or M + 1 and M + 18 respectively.

## 6.2  Instruments

### Single focusing mass spectrometer

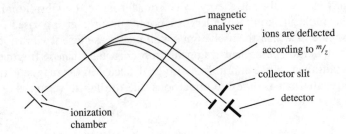

**Fig. 6.6**  Schematic of a single focusing mass spectrometer.

In this simplest arrangement (Fig. 6.6), samples having an appreciable volatility are introduced by connecting an ampoule to an inlet reservoir. Less volatile materials are introduced directly into the ionization chamber on the end of a probe which can be heated. Electron impact ionization is carried out using a beam of high energy electrons (*ca* 70 eV). Ionization is possible with electrons of about 10 eV and the excess energy is transferred to vibrational modes of the molecule and results in fragmentation of the ionized species. The positively charged species so produced are accelerated under a 4–8kV potential and focused into a tight beam. The collimated beam is then passed through a strong magnetic field placed at right angles to the line of flight of the ions. Lighter ions are deflected more than heavier ions, the separated ion beams are individually detected and the intensity of each beam measured (*relative abundance*). The ion beams are focused onto the detector by varying the magnetic field strength; sweeping the magnetic field permits a mass range to be surveyed.

### Double focusing mass spectrometer

This instrument has an electrostatic analyser placed in series with the

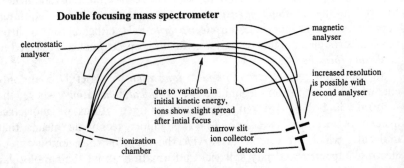

**Fig. 6.7**  Schematic of a Nier Johnson arrangement double focusing mass spectrometer.

magnetic analyser (Fig. 6.7). The electrostatic analyser refocuses the small energy spread to be found within any collection of ions having the same mass. The resolution of the instrument is enhanced still further by using narrower slits before the collector. Although this has the effect of reducing sensitivity, it means that molecular masses can be measured with an accuracy of 1–2 ppm. Consequently it is possible to determine molecular composition without the necessity for combustion analysis, permitting distinction between two molecules having the same integral molecular mass but different composition (Fig. 6.8).

## Quadrupole mass spectrometer

This arrangement uses four voltage carrying rods about 10–30 cm long (the **quadrupole**) running the length of the flight path of the ions. Ions travel with constant velocity parallel to the rods but, as a result of the application of direct current and radio-frequency voltages to the rods they acquire complex oscillations in their path (Fig. 6.9). For any $m/z$ ratio there is a **stable oscillation** which permits ions to traverse the whole length of the analyser without hitting the rods and being lost. In other words, under any set of applied dc and rf voltages, only ions of a particular $m/z$ ratio will be collected and analysed. Mass scanning is carried out by varying the two applied voltages while keeping their ratios constant.

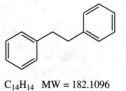

$C_{14}H_{14}$  MW = 182.1096

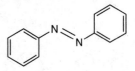

$C_{12}H_{10}N_2$  MW = 182.0844

**Fig. 6.8**  The precision possible with high resolution mass spectrometry permits molecular composition to be deduced.

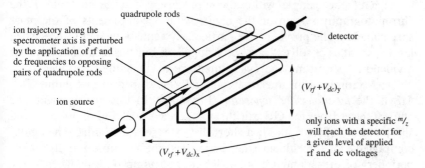

**Fig. 6.9**  Schematic of the arrangement of the quadrupole mass spectrometer.

## Time-of-flight mass spectrometer

This type of spectrometer works on the principle that all singly charged species subjected to the same potential difference will achieve the same translational energy as measured in electron volts. However, lighter

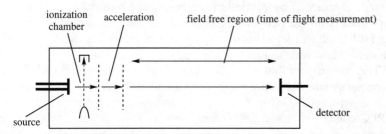

**Fig. 6.10**  Schematic of a time of flight mass spectrometer (Adapted and reproduced with permission of VG Organic).

particles will be accelerated faster than heavier ones and will thus have a shorter ***time-of-flight*** over a particular distance. The particles are allowed to pass through a field-free region and are analysed by their arrival time at the detector (Fig. 6.10). As the time differences are in the order of $10^{-7}$ s, fast electronics are necessary for adequate resolution, but it is difficult to achieve better than about 2 mass units.

## On-line chromatography–mass spectrometry (GC–MS and HPLC–MS)

Coupling a mass spectrometer to the effluent of a gas or liquid chromatograph combines the powerful analytical capability of mass spectrometry with the high degree of separation possible with gas chromatography (usually capillary gas chromatography) or high performance liquid chromatography. Commercial instruments commonly use a spectrometer based upon a quadrupole mass filter due to the ruggedness of this particular design.

With on-line chromatography-mass spectrometry, it is necessary to separate the molecules to be analysed from the carrier gas or elution solvents—the chromatography column is at a positive pressure; whereas the mass spectrometer operates at $10^{-6}$ torr. Thus the crucial part of such a set-up is the ***interface*** between the two instruments which must permit passage of the sample without compromising the throughput of the chromatography system or the high vacuum requirements of the mass spectrometer. The problems are particularly acute with HPLC–MS when large quantities of solvent must be stripped off but this technique pays the dividend of permitting analysis of mixtures of involatile components.

One example of an interface commonly used in packed column GC–MS is the ***jet molecular separator*** (Fig. 6.11). In this arrangement, the effluent gases from the GC exit through a fine outlet and impinge onto a sharp edged collector nozzle a short distance from the outlet. The lighter components of the effluent (the carrier gas) diffuse more rapidly to the periphery of the jet and are not collected, but pumped away; whereas the heavier components selectively pass into the spectrometer to be analysed.

It is true to say that the scale of problems encountered with HPLC–MS means that there is no universal interface for all situations. The first to become commercially available was the ***moving belt interface*** which, as its name implies, involves deposition of the HPLC effluent onto an endless belt relaying sample to the ionization chamber through a series of vacuum chambers to remove the majority of the solvent (Fig. 6.12). Residual solvent is removed by gentle heating before the residual sample is flash-vaporized from the belt surface into the ionization source. On return to the HPLC effluent jet to receive more sample, the belt is cleaned to avoid contamination with residues of the previous deposition.

The moving belt interface has now been largely superseded by the ***thermospray interface*** which is especially useful for analysis of polar materials with aqueous eluents (Fig. 6.13). In this arrangement, the effluent from the HPLC column is heated and forced through a fine jet into an evacuated chamber to give an aerosol. This results in simultaneous volatilization and ionization of the droplets by buffer (commonly ammonium acetate) present

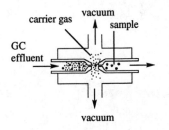

**Fig. 6.11**  Schematic of a jet molecular separator.

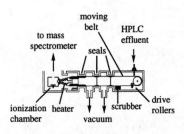

**Fig. 6.12**  Schematic of a moving belt interface (Adapted and reproduced with permission of VG Organic).

in the mobile phase. Statistically, some droplets develop an excess positive charge and some an excess negative charge. Solvent is pumped off concentrating the charges onto the sample and the positively ionized components directed into the mass spectrometer by the repelling electrode.

## 6.3 Ionization techniques

### Electron impact (EI)

In this conceptually simplest technique, the sample is bombarded with a beam of high energy (*ca* 70 eV) electrons which, on impact, transfer about 20 eV of energy to the molecule leading to ejection of electrons to give positively charge species (Fig. 6.14). As mass spectrometers are designed to record species possessing a single positive charge, the largest fragment corresponds to the radical cation $M^{\bullet+}$ and gives the molecular weight of the substance.

As ionization requires about 15 eV energy, the 5 eV or so of excess

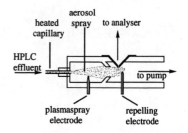

**Fig. 6.13** Schematic of a thermospray interface (Adapted and reproduced with permission of VG Organic).

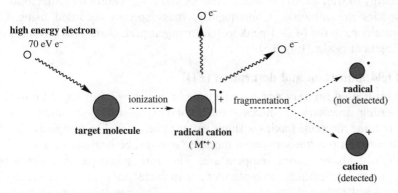

**Fig. 6.14** Electron Impact ionization.

energy transferred by the impacting electron passes into the vibrational modes of the radical cation. This causes the radical cation to fragment giving smaller species and those retaining a charge may also be analysed and detected.

### Chemical ionization (CI)

Many compounds are too labile to stand up to impact by 70 eV electrons without total fragmentation Consequently, analysis of such materials using EI would result in no molecular ion being detected. In the chemical ionization procedure, the sample is introduced into the ionization chamber at about 1 torr pressure together with a carrier gas such as hydrogen, ammonia or methane. It is the carrier gas which is ionized selectively by the high energy electrons, as it is present in excess. The initial radical cations (***primary ions***) react with further neutral molecules of the carrier gas to give ***secondary ions***. It is these secondary ions which, having a longer lifetime, can react with molecules of the sample to give ***quasi-molecular ions***. The scheme of Fig. 6.15 shows the overall process using ammonia as the ionizing reagent, as it is commonly used for CI.

Thus, with CI analysis, M + 1 and M + RH peaks (where R = the reagent gas) are recorded, with the M + 1 peak usually being the

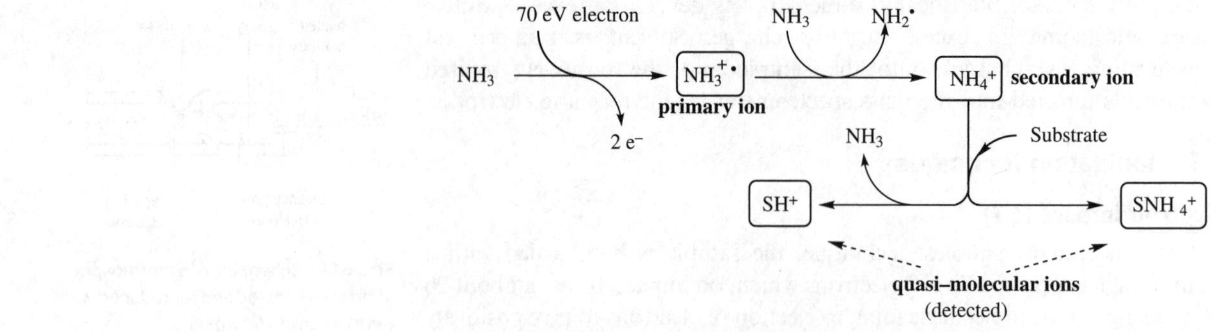

**Fig. 6.15**   The chemical ionization pathway.

dominant peak. As the ions have been produced by a chemical reaction and not by direct electron impact, they do not have a great excess of energy to transfer to vibrational modes and fragmentations of the ionized species are minimal. Consequently, mass spectra recorded using CI usually have the M + 1 peak as the strongest peak (***base peak***) with fewer fragment peaks (Fig. 6.5).

### Field ionization and desorption (FD)

This is used for very non-volatile samples which will not stand up to the heating necessary to increase their volatility (mass spectrometry relies upon the molecule having a slight vapour pressure). The sample is placed directly onto a fine wire on a modified inlet probe held at $+8$ kV at, or slightly above, room temperature. The wire possesses fine dendritic growths along which the molecules are subjected to local, extremely high, electrical potentials of up to $10^{10}$ V m$^{-1}$. This results in abstraction of electrons from the molecules into unfilled orbitals of the metal and the positive ions so produced are immediately expelled from the probe by Coulombic repulsion without the need for excessive external heating. In this manner, the ions produced have little excess energy and spectra recorded using this technique usually only show the molecular ion or $MH^{+}$ and sometimes $MNa^{+}$ peaks due to intermolecular interactions occurring close to the surface of the probe. The technique can be used to analyse molecules having molecular weights up to about 10 k dalton.

### Secondary ion mass spectrometry (SIMS) and fast atom bombardment (FAB)

These are mild ionizing techniques for large polar molecules such as peptides and are based upon an indirect ionization approach. In SIMS, argon is first ionized by electron impact and the beam of $Ar^{\bullet+}$ directed towards the sample causing ionization on collision. In FAB, the substrate is dispersed in an involatile matrix such as glycerol and is ionized by impact with high energy xenon atoms obtained by passing $Xe^{\bullet+}$ ions, accelerated to about 10 keV, through more xenon gas (Fig. 6.16). The xenon radical cations pick up an electron, becoming xenon atoms with high translational energy, while the $Xe^{+}$ ions also formed are removed electrostatically. The ***fast atoms*** so produced are allowed to impact on the sample to be analysed.

Ionization occurs by transfer of translational energy from the accelerated xenon atoms, with minimal transfer to vibrational modes of the substrate and hence reduced fragmentation of the ionized molecules. The relatively polar medium of the glycerol acts as an ionizing solvent and favours ion formation. In addition, as the ions take some time to diffuse to the surface, the ion stream may last for up to 30 minutes as opposed to about 2 seconds for normal ionization techniques. Usually $MH^+$ is seen together with $MNa^+$ and $MK^+$ from inorganic impurities in the matrix. Protonated glycerol derived ions dominate the spectrum at $m/z$ $92n + 1$ corresponding to $(C_3H_8O_3)_nH^+$, but these ions do not interfere for the types of molecule studied with this technique. Fragment ions are useful, particularly when the technique is used for protein analysis as the fragments may be used to sequence the peptide chain.

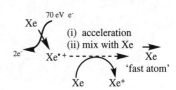

**Fig. 6.16**  Generation of 'fast' xenon atoms.

### Miscellaneous desorption techniques

Several techniques have been developed to deliver a rapid pulse of energy to the sample which results mainly in breaking intermolecular interactions such as hydrogen bonds rather than cleaving covalent bonds within the molecule. The energy may be delivered by impact of extremely energetic ionized particles such as $^{142}Ba^{18+}$ and $^{106}Te^{22+}$ resulting from fission of $^{252}Cf$ (*plasma desorption*) or a very finely focused laser beam (*laser desorption*, particularly *matrix assisted laser desorption*) and causes simultaneous vaporization and ionization of the material. These procedures permit molecules having masses up to 25 k dalton to be analysed.

### Electrospray

Macromolecules, such as proteins, would be decomposed by any attempt to volatilize or ionize them under usual conditions. Electrospray enables molecules to be taken directly from a solution to the gas phase ionized state by passing the solution through the exit of a fine needle held at an electrical potential around 4 kV (Fig. 6.17). The solution disperses into a fine mist of ionized droplets which quickly lose their solvent, leaving behind an aerosol of protonated sample which is then desorbed into the gas phase and interfaced with the vacuum for mass spectrometric analysis. Fragmentation is not observed under such mild conditions and it is possible that even the tertiary structures of some proteins are retained. In principle, the mass range available by such a technique is limited only by the capabilities of the instrument and proteins up to 150 k dalton may be analysed fairly routinely.

As each molecule of the substrate may possess a variable number of protons, a series of $m/z$ peaks is recorded and from this the true mass may be obtained by computer analysis. This technique is probably the 'softest' ionization procedure available and permits molecular weight determinations of large molecules with accuracies of better than 0.01%.

## 6.4    Interpretation of spectra

### The molecular ion

As already discussed, the primary information given by mass spectrometry is the molecular weight of the substance as the spectrometer

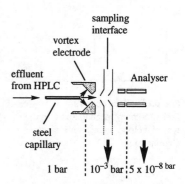

**Fig. 6.17**  Schematic of an electrospray source (Adapted and reproduced with permission of VG Organic).

measures $m/z$. However, deciding which peak actually is the molecular ion is not necessarily straightforward and depends upon the ionization technique. Only EI leads to $M^{\bullet+}$ being recorded; if softer ionization techniques such as CI, FAB or electrospray are being used we measure $MH^+$, $MNH_4^+$ or $MNa^+$ depending upon the technique used (Fig. 6.5a).

Mass spectrometry measures individual atoms and so this enables the isotopic constitution of the sample to be measured. While very few elements (F, I, P) are truly monoisotopic, many other elements (H, N) of interest to the organic chemist can be considered to be effectively monoisotopic, due to the low abundance of other isotopes. However, there are some elements (C, Br, Cl, S, Si) of which the isotopic composition has an important bearing on the peaks observed around the molecular ion.

Carbon is roughly 99% $^{12}C$ and 1% $^{13}C$. Consequently, in a molecule containing 10 carbon atoms, there is statistically a 10% chance that one of the atoms will be $^{13}C$. Therefore, the $M^{\bullet+}$ peak will be accompanied by a peak at $M+1$ having 10% intensity due to the those molecules containing one $^{13}C$ atom. For such a molecule the peak at $M+2$ is usually too small to be detected and does not cause confusion, but it is important to be aware that moderately sized molecules containing C, H, N and O have a molecular ion which is the major peak within the cluster of the peaks found at the highest mass in the spectrum. For larger molecules this identification problem becomes more complicated; although, for molecules with molecular weights greater than 10 k dalton the level of imprecision is of lesser importance.

Two important situations arise with bromine and chlorine. Bromine consists of two isotopes, $^{79}Br$ and $^{81}Br$, in approximately 1 : 1 ratio. Molecules containing one bromine atom show a molecular ion which consists of two peaks of roughly equal intensity separated by two mass units. Patterns resulting from two and three bromine atoms in a molecule are shown in Fig. 6.18. The relative intensities of peaks for such di-isotopic systems can be calculated using the binomial expansion $(a + b)^n$. Chlorine has two isotopes $^{35}C$ and $^{37}C$ in about a 3 : 1 ratio. Molecules containing a chlorine atom show a molecular ion consisting of 2 peaks separated by two mass units with the higher peak about one third the intensity of the lower mass peak (Fig. 6.19 and 6.20).

There is always the possibility that the molecule is too unstable if the analysis has been carried out under EI ionization conditions and never gives a molecular ion, the highest peak observed being due to a fragment (Fig. 6.5b). Suspect this particularly if there are fragment ions having between 3–14 mass units less than the apparent 'molecular ion'. Commonly encountered organic molecules are not capable of readily losing fragments within this range and what is in fact being observed are different sets of fragment ions arising from an unobserved molecule which has undergone total fragmentation before analysis.

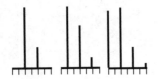

**Fig. 6.18**   Isotope patterns of Br, Br₂, Br₃ species.

**Fig. 6.19**   Isotope patterns of Cl, Cl₂, Cl₃ species.

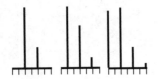

156

158

**Fig. 6.20**   Mass spectrum of 2-chloroethoxybenzene showing the typical isotope pattern for a single chlorine atom.

## 6.5   The 'nitrogen rule'

Nitrogen is the only commonly encountered odd valency element possessing an even atomic weight in organic molecules. Any compound containing one or an odd number of nitrogen atoms will have an odd molecular weight; conversely, compounds lacking nitrogen or containing even numbers of nitrogens have an even molecular weight. An odd molecular ion in a spectrum obtained under conditions of electron impact ionization should therefore alert you to the fact that the substance possibly contains nitrogen but don't forget that it might simply be that you are not observing the molecular ion but a fragment peak instead. Of course, the inverse applies to spectra obtained under chemical ionization conditions if we consider $MH^+$.

## 6.6   Fragmentation pathways

The occurrence of fragment ions within the mass spectrum provides useful information about the constitution of the molecule as fragmentations often occur at labile bonds or to give particularly stable fragments. Fragmentation may involve cleavage of a single bond or cleavage of two bonds.

### One-bond cleavages

Radical cations may fragment by either heterolytic bond cleavage to give a cation and a radical or by homolytic cleavage to generate a new radical cation.

$$X-Y^{\bullet +} \xrightarrow{\text{heterolytic cleavage}} \underset{\text{detected}}{X^+} + Y^\bullet \quad \text{or} \quad X^\bullet + \underset{\text{detected}}{Y^+}$$

$$X-Y^{\bullet +} \xrightarrow{\text{homolytic cleavage}} \underset{\text{detected}}{X^{\bullet +}} + Y \quad \text{or} \quad X + \underset{\text{detected}}{Y^{\bullet +}}$$

$$X\overset{\frown}{\underset{\bullet}{-}}Y \longrightarrow X^+ \quad Y^-$$

*Heterolytic* bond cleavage involves movement of a pair of electrons.

On the other hand, cations only degrade by a heterolytic process to generate a new cation and a neutral species, as the homolytic process to yield a radical and a radical cation is energetically unfavourable.

$$X-Y^+ \xrightarrow{\text{heterolytic cleavage}} \underset{\text{detected}}{X^+} + Y \quad \text{or} \quad X + \underset{\text{detected}}{Y^+}$$

$$X\overset{\frown}{\underset{\smile}{-}}Y \longrightarrow X^\bullet \quad Y^\bullet$$

*Homolytic* cleavage involves movement of single electrons.

This leads to the ***even electron rule***, a guideline which states that even-electron species do not fragment to odd-electron species. In other words, a cation will not fragment to form a radical and a radical cation. One-bond cleavage gives fragments which are valence unsatisfied. Compounds or fragments with an even molecular weight give one-bond cleavage fragments which are odd and *vice versa*.

### Two-bond cleavages

These cleavages result from chemical reactions of the radical cation. Even molecular weight compounds give even molecular weight two-bond cleavage fragments and *vice versa*, again making these cleavages easy to identify.

Only one of the fragments retains the charge and is detected. Which fragment this will be can be predicted by the **Stephenson–Audier rule** which states that the positive charge will reside on the fragment with the lowest ionization potential. This can be approximated by deciding which fragment has the greater potential for resonance stabilization (Fig. 6.21).

Note that one-bond-cleavage of an even molecular weight precursor leads to an odd molecular weight fragment. The charge is associated with the fragment with the better resonance stabilization.

Conversely, two-bond-cleavage of an even molecular weight precursor leads to even molecular weight fragments.

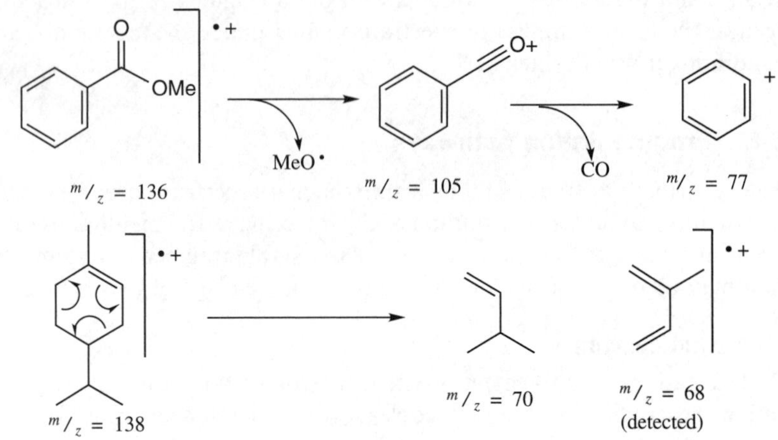

$m/_z = 136$    $m/_z = 105$    $m/_z = 77$

$m/_z = 138$    $m/_z = 70$    $m/_z = 68$ (detected)

**Fig. 6.21**    The Stephenson–Audier rule predicts that the charge resides on the fragment with the lowest ionization potential.

## Metastable peaks

These are a result of fragmentations occurring in the *field free region* of the spectrometer, which is the area outside the ionization chamber but before the magnetic analyser, and means that fragmentation occurs after some initial acceleration of the original charged fragment has already taken place. The excess momentum is passed on to the daughter ions and means that they possess a higher velocity than they would have achieved if they had been formed before acceleration. They are not shown on 'bar' spectra (which only give integral peaks) but can be seen on photographically recorded spectra. They are broad and occur at non-integral positions in the spectrum. The relationship between the daughter and parent ions is expressed by the equation shown in the margin.

$$m^* = \frac{(m_2)^2}{m_1}$$
$m_1$ = parent ion mass
$m_2$ = daughter ion mass
$m^*$ = metastable peak mass

Metastable peaks are relatively uncommon, as their observation requires fragmentation to occur in the field-free region between the ionization chamber and the magnetic analyser. Any faster and the daughter ion is formed in the ionization chamber to give the normal fragment; any slower and the fragmentation occurs in the analyser and the fragment does not reach the detector. This requires fragmentations occurring with lifetimes of around $10^{-4} \rightarrow 10^{-6}$s. Of course, merely knowing the mass of the metastable peak ($m^*$) is insufficient to arrive at unambiguous values for the parent and daughter ions $m_1$ and $m_2$. However it is possible to obtain the masses of the parent and daughter peaks by inspection as they usually occur as strong peaks in the spectrum. Alternatively, computer analysis of spectra makes the calculation simple.

## 6.7 Common fragmentation pathways

Fragmentations are particularly favoured at weak bonds or to produce stable fragments and you should refer to Appendix 4 for lists of common fragmentations and fragment ions. Usually it is obvious when simple substituents such as alkyl groups are lost, but there are several important cleavages which should be remembered.

### Tropylium ion

Organic substrates containing a benzyl group show a strong peak, usually the base peak at $m/z = 91$ due to a one-bond cleavage. This is normally accompanied by a second peak at $m/z = 65$ resulting from loss of ethene in a two-bond cleavage step.

### Acylium ions

Compounds possessing acyl groups give strong fragment ions corresponding to the acylium cations. For example acetates and methyl ketones give a strong peak at $m/z = 43$ and benzoates and phenyl ketones a strong peak at $m/z = 105$. These species are usually accompanied by peaks corresponding to subsequent extrusion of carbon monoxide (Fig. 6.21).

### Favourable bond cleavages

*Reverse Diels–Alder.* This commonly occurs with cyclohexene derivatives and the charged fragment is usually the diene.

The reverse Diels–Alder reaction.

*McLafferty rearrangement.* This is a reverse ene reaction and is observed in mass spectra of carbonyl compounds possessing a γ-hydrogen. Depending upon the type of substrate (ketone, aldehyde or ester), the charge can be found on either fragment but usually it remains on the carbonyl derived fragment.

The McLafferty rearrangement.

A similar fragmentation can occur with alkenes possessing a γ-hydrogen

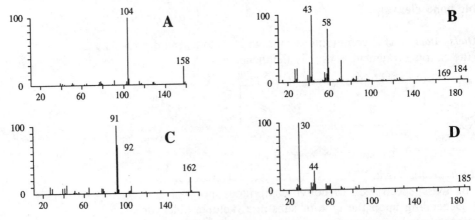

## Loss of small neutral molecules

Generally, elimination of small, neutral molecules occurs readily. In particular, alcohols lose water to give M–18, acetates lose acetic acid to give M–60 and halides lose HX. In this latter case, it will be observed that the characteristic isotope pattern of the molecular ion due to the presence of chlorine or bromine disappears in the fragments which are halogen free.

## 6.8   Exercises

*Note.*   All spectra were recorded under EI conditions and molecular ions are recorded in each case.

1. The mass spectrum of 2-butenal gives a fragment ion at $m/z$ 69 (M$^+$ − 1) which shows an intensity of 29% compared with the base peak. Suggest a fragmentation pathway to account for this peak and suggest why the fragment should be stabilized.

2. The mass spectrum of 3-butyn-2-ol shows a base peak at $m/z$ 55. Explain why the fragment giving rise to this peak should be stable.

3. Below are spectra of dodecylamine, 2-dodecanone, 1-phenylhexane and 4-phenylcyclohexene. Assign each spectrum explaining the salient features.

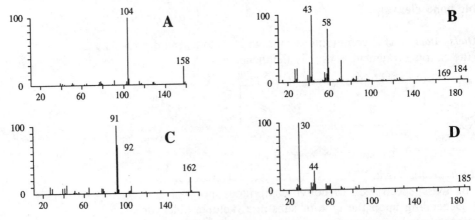

4. The following are mass spectra of three isomeric alcohols, 3-pentanol, 2-pentanol and 3-methyl-1-butanol. Assign each spectrum with reasoning.

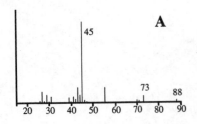
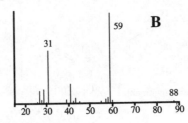
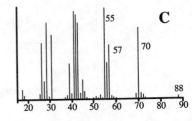

5. Mass spectrometric analysis of compounds A–C gave the spectra shown. Compounds A and B gave precipitates when treated with alcoholic silver nitrate, that from A being creamish-yellow and that from B being white. Compounds A and B may are converted into C on treatment with hot aqueous sodium hydroxide. Deduce the structures of A–C.

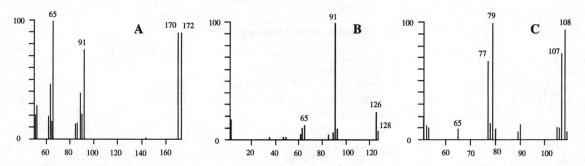

## Further reading

1. For texts dealing with data interpretation see: F. W. McLafferty and F. Turecek, *Interpretation of Mass Spectra*, 4th edn., 1993, University Science Books, Mill Valley, Ca.; D. H. Williams and I. Fleming, *Spectroscopic Methods in Organic Chemistry*, 5th edn. 1995, Chapter 4, McGraw–Hill, London. J. H. Benyon, R. A. Saunders and A. E. Williams, *The Mass Spectra of Organic Molecules*, 1968, Elsevier, Amsterdam.

2. For experimental aspects of the technique see: J. R. Chapman, *Practical Organic Mass Spectrometry*, 1985, Wiley–Interscience, Chichester; F. W. McLafferty (ed.), *Tandem Mass Spectrometry*, 1983, Wiley–Interscience, New York.

3. For a discussion of electrospray mass spectrometry see: M. Mann and M. Wilm, *TIBS*, 1995, June, 219.

4. Recent catalogues of data include: *Eight Peak Index of Mass Spectra*, 3 volumes, 4th edn., 1991, Royal Society of Chemistry, Cambridge; F. W. McLafferty and D. B. Stauffer (eds.), *The Important Peak Index of the Registry of Mass Spectral Data*, 3 vols., 1991, Wiley–Interscience, New York; F. W. McLafferty and D. B. Stauffer (eds.), *The Wiley/NBS Register of Mass Spectral Data*, 6 vols., 1989, Wiley–Interscience, New York; K. Pfleger, H. H. Maurer and A. Weber (eds.), *Mass Spectra and G. C. Data of Drugs, Poisons, Pollutants and their Metabolites*, 3 vols., 1992, VCH, Weinheim.

# Exercise answers

## 1.6 Introductory theory

1. frequency $= 2 \times 10^{13}$ s$^{-1}$
2. wavenumber $= 667$ cm$^{-1}$.
3. 1750 cm$^{-1}$: $\lambda = 5.7$ $\mu$m, $v = 5.25 \times 10^{13}$ s$^{-1}$
   3500 cm$^{-1}$ : $\lambda = 2.9$ $\mu$m, $v = 1.05 \times 10^{12}$ s$^{-1}$
4. $v = 1.2 \times 10^{15}$ s$^{-1}$
   a single molecule absorbs $7.8 \times 10^{-19}$ J
   1 mole would absorb 469 kJmol$^{-1}$.
5. wavelength 314 nm,
   frequency $9.56 \times 10^{14}$, s$^{-1}$, ultra-violet

## 2.3 Ultraviolet–visible spectroscopy

1. $\varepsilon = 14$, forbidden transition, n $\rightarrow$ $\pi^*$
2. The two possible transitions correspond to the n $\rightarrow$ $\pi^*$ (320 nm, C=O) and
   $\pi \rightarrow \sigma^*$ (213 nm, C=C, strongest)
3. The system has a conjugated double bond.

4. base value (217) + 3 $\times$ alkyl groups (3 $\times$ 5) = 232
   base value (215) + $\alpha$-alkyl (10) + 2 $\times$ $\beta$-alkyl (2 $\times$ 12) = 249
   base value (CO$_2$Me, 230) + *meta* Br (2) = 232
   base value (CHO, 250) + *meta* Cl (0) + *meta* OMe (7) = 257
5. $\lambda_{max}$ (a) 355 nm      (b) 240 nm      (c) 289 nm
6. $\varepsilon$ 16538 (257)   $\varepsilon(\lambda)$ 12032 (252)    1604 (282)    1020 (292)   $\varepsilon(\lambda)$ 13500 (252)

## 3.6 Infrared spectroscopy

1. (a) benzyl alcohol: 3300 cm$^{-1}$ (OH stretch), 3000 (C–H stretch), 2000
   (aromatic overtones), 1450 (weak aromatic ring C=C stretch), 1100 (C–O
   stretch), 750 and 700 (monosubstituted aromatic)
   (b) 1-octene: 3000 (C–H stretch), 1650 (C=C stretch), 1500 (C–H bend)
2. (a) acid: 3500 cm$^{-1}$ (broad OH stretch), 1710 (C=O), 1150 (C–O)
   (b) acyl chloride: 1805 (C=O)
   (c) aldehyde: 2700 (C–H of aldehyde), 1730 (C=O)
   (d) amide: 1658 (amide I, C=O stretch), 1548 (amide II, N–H)
3. (a) methyl phenyl ketone: 1683 cm$^{-1}$ (C=O lowered due to conjugation)
   (b) butyrolactone: 1771 (C=O lactone), 1180 (C–O stretch)
   (c) ethyl ethanoate: 1745 (C=O), 1250 (C–O stretch)
   (d) cyclohexenone: 1713 (C=O)

## 4.7  NMR spectroscopy: the basics

1.

2.  1c, 2a, 3b.
3.  The molecule is Penten-3-ol.
4.  A dynamic equilibrium exists due to restricted rotation about the amide bond (see Section 4.6). $k_c = 88.9 \text{ s}^{-1}$, $\Delta G^{\ddagger} = 74.6 \text{ KJmol}^{-1}$.

## 5.6  NMR spectroscopy: further topics

1.

2.

3.

## 6.8  Mass spectrometry

1.  The fragment ion at *m/z* 69 corresponds to loss of H. The fragment is stabilized by conjugation

2.  The base peak at *m/z* 55 is due to loss of methyl. The fragment is stabilized by resonance and conjugation

3.  (a) 4-phenylcyclohexene ($C_{12}H_{14}$, MW 158)

(b) 2-dodecanone ($C_{12}H_{24}O$, MW 184)

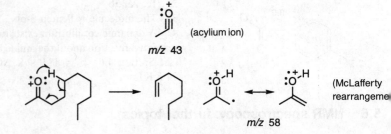

(acylium ion)

*m/z* 43

(McLafferty rearrangeme|

*m/z* 58

(c) 1-phenylhexane ($C_{12}H_{18}$, MW 162)

*m/z* 162     $-C_5H_{11}^{\bullet}$

(tropyliun|

*m/z* 91

(reverse ene)

*m/z* 92

(d) dodecylamine ($C_{12}H_{27}N$, MW 185)

$H_2C\overset{+}{=}\overset{\cdot\cdot}{N}H_2$   *m/z* 30

$-\overset{\cdot\cdot}{N}H_2$   *m/z* 44

4.  (a) 2-pentanol

*m/z* 88     *m/z* 73     *m/z* 45

(b) 3-pentanol

*m/z* 88     *m/z* 59     *m/z* 31

(c) 3-methyl-1-butanol

*m/z* 88     *m/z* 70     *m/z* 57

5. Molecule A is benzyl bromide: A creamish-yellow precipitate with silver nitrate indicating bromide. The mass spectrum shows two equal height peaks at 170, 172 corresponding respectively to $^{81}Br$ and $^{79}Br$. Subtracting the mass of bromine gives a mass of 91. This is present along with a peak at 65 which is indicative of the benzyl group. (For fragmentation pattern see Section 6.7 – tropylium ion).

Molecule B is benzyl chloride: A white precipitate with silver nitrate indicates chloride. The mass spectrum peaks of 128 and 126 are in approximately a 3:1 ratio indicating $^{37}Cl$ and $^{35}Cl$. Subtracting the mass of chlorine gives 91.

Molecule C is benzyl alcohol: as halides can be converted by treatment with sodium hydroxide to alcohols.

# Appendices

## Appendix 1   UV-vis spectra predictive rules for extended conjugation

The following tables give the position of $\lambda_{max}$ of the lowest energy $\pi \rightarrow \pi^*$ transition

### Conjugated dienes

Example

| | wavelength (nm) |
|---|---|
| Heteroannular or open chain diene | 214 |
| Homoannular diene | 253 |
| **Increment for:** | |
| (a) Every alkyl substituent or ring residue | 5 |
| (b) The exocyclic nature of any double bond | 5 |
| (c) Each double bond extension on the diene | 30 |
| (d) Auxochrome | |

| | | | |
|---|---|---|---|
| −Oacyl | 0 | −SR | 30 |
| −OR | 6 | −NR$_2$ | 60 |

| | | |
|---|---|---|
| Parent homoannular diene | 253 | |
| Alkyl substituents (×2) | 10 | |
| Ring residues (×2) | 10 | |
| TOTAL | 273 nm | |
| Observed value | 272 nm | |

### $\alpha$, $\beta$-Unsaturated carbonyl compounds

The following values apply to EtOH and MeOH solutions only. Other solvents require a correction: CHCl$_3$ +1 nm, Et$_2$O +7 nm, C$_6$H$_{14}$ +11 nm, H$_2$O −8 nm

Example

| | | wavelength (nm) | | |
|---|---|---|---|---|
| $\alpha$, $\beta$-Unsaturated 6-ring or acyclic ketone | | 215 | | |
| $\alpha$, $\beta$-Unsaturated 5-ring ketone | | 202 | | |
| $\alpha$, $\beta$-Unsaturated aldehyde | | 207 | | |
| **Increment for:** | | | | |
| (a) Every double bond extending conjugation | | 30 | | |
| (b) Every alkyl group or ring residue | $\alpha$ | 10 | | |
| | $\beta$ | 12 | | |
| | $\gamma$ and higher | 18 | | |
| (c) The exocyclic nature of any double bond | | 5 | | |
| (d) Homodiene component | | 39 | | |
| (e) Auxochrome | | | | |
| −OH | $\alpha$ | 35 | −Oacyl  $\alpha, \beta, \delta$ | 6 |
| | $\beta$ | 30 | −SR  $\beta$ | 85 |
| | $\delta$ | 50 | −Cl  $\alpha$ | 5 |
| −OR | $\alpha$ | 5 | $\beta$ | 12 |
| | $\beta$ | 0 | −Br  $\alpha$ | 25 |
| | $\gamma$ | 17 | $\beta$ | 30 |
| | $\delta$ | 31 | −NR$_2$  $\beta$ | 95 |

| | |
|---|---|
| Parent $\alpha$, $\beta$-unsaturated 6-ring ketone | 215 |
| $\alpha$–Ring residue | 10 |
| $\beta$–Ring residue | 12 |
| $\beta$–Alkyl group | 12 |
| TOTAL | 249 nm |
| Observed value (EtOH) | 243 nm |

### Substituted benzene derivatives Ar–COX

Example

| | | | wavelength (nm) | | |
|---|---|---|---|---|---|
| **Value for parent chromophore:** | | | | | |
| | −X = alkyl or ring residue | | 246 | | |
| | −X = H | | 250 | | |
| | −X = OH or −OR | | 230 | | |
| **Increment for:** | | | | | |
| alkyl or ring residue | ortho, meta | 3 − | Br | ortho, meta | 2 |
| | para | 0 | | para | 15 |
| −OH, −OR | ortho, meta | 7 | −NH$_2$ | ortho, meta | 13 |
| | para | 25 | | para | 58 |
| −O$^-$ | ortho | 11 | −NHacyl | ortho, meta | 20 |
| | meta | 20 | | para | 45 |
| | para | 78 | −NHR | ortho, meta | 73 |
| −Cl | ortho, meta | 0 | −NR$_2$ | ortho, meta | 120 |
| | para | 10 | | para | 85 |

| | |
|---|---|
| Parent chromophore | 246 |
| Alkyl residue ortho– | 3 |
| Methoxy–residue ortho | 7 |
| Methoxy–residue para– | 25 |
| TOTAL | 281 nm |
| Observed value | 275 nm |

## Appendix 2   Infrared correlation tables

For a more exhaustive series of tables see: D. H. Williams and I. Fleming, *Spectroscopic Methods in Organic Chemistry*, 5th edn, McGraw–Hill, London, 1995, Chapter 2, pp 34–57.

| $\nu$ (cm$^{-1}$) | Functionality | Comment |
|---|---|---|
| **4000–2500 region** | | |
| 3600 | O–H | Free non H–bonded is sharp |
| 3500–3000 | O–H | H–bonded, broad |
| | N–H | Usually broad, amine or amide |
| 3300 | ≡C–H | Sharp and strong |
| 3100–2700 | C–H | Variable, usually strong for sp$^3$, weak for sp$^2$ hybridized C–H |
| 3500–2500 | CO$_2$H | Broad, H–bonded carboxylic acid OH |
| **2500–1900 region** | | |
| 2350 | CO$_2$ | Path length imbalance not sample |
| 2200 | C≡C, C≡N | Usually weak. If a terminal alkyne, shows a peak at 3300 cm$^1$ |
| 2200–1900 | X=Y=Z | Strong, allene, isocyanate, azide, diazo |
| **1900–1500 region** | | |
| 1850–1650 | C=O | Strong $\alpha$, $\beta$-unsaturation 1650–1690 cm$^{-1}$, small ring ketoness ($\leq$ 5 membered) and $\alpha$-electron withdrawing groups (eg halogen) 1750–1850 cm$^{-1}$ |
| 1650–1500 | C=C, C=N | Usually weak. Stronger if conjugated. Absent for symmetrical alkenes |
| 1600 | C=C (arom) | Variable. Usually associated with peaks in the fingerprint region. |
| 1550 | NO$_2$ | Strong |
| **1500–600 (Fingerprint) region** | | |
| 1350 | NO$_2$ | Strong |
| | –SO$_2$– | Strong, also a peak at 1150 cm$^{-1}$ |
| 1300–250 | P=O | Strong |
| 1300–1000 | C–O | Strong, alcohol, ether, ester. |
| 1150 | –SO$_2$– | Strong, see above |
| 850–700 | C–H (arom) | Indicates substitution pattern (p. 33) |
| 800–700 | C–Cl | Strong, obscured by CHCl$_3$ |

### Appendix 3   NMR correlation tables

Tables of $^1$H shifts are adapted from E. Pretsch, T. Clerc, J. Seibl, and W. Simon. *Tables of Spectral Data for the Structure Determination of Organic Compounds*, Springer-Verlag, with permission.

#### Tables for estimating $^1$H chemical shifts of CH$_2$ and CH groups

$\delta$ CH$_2$X$_1$X$_2$ = 1.25 + $\Delta_1$ + $\Delta$2 ppm   $\delta$ CHX$_1$X$_2$X$_3$ = 1.5 + $\Delta_1$ + $\Delta_2$ + $\Delta_3$ ppm

| X | $\Delta$ | X | $\Delta$ | X | $\Delta$ |
|---|---|---|---|---|---|
| –Alkyl | 0.0 | –OR | 1.5 | –CN | 1.2 |
| –C=C | 0.8 | –OH | 1.7 | –CO-R | 1.2 |
| –C≡C | 0.9 | –OAr | 2.3 | –CO-OR | 0.7 |
| –Aryl | 1.3 | –OCO-R | 2.7 | –CO-OH | 0.8 |
| –SR | 1.0 | –OCO-Ar | 2.9 | –I | 1.4 |
| -NR$_2$ | 1.0 | –CHO | 1.2 | –Br | 1.9 |
| –NO$_2$ | 3.0 | | | –Cl | 2.0 |

## Tables for estimating $^1$H chemical shifts of alkene groups

$\delta\,CH{=}C = 5.25 + \Delta_{gem} + \Delta_{cis} + \Delta_{trans}$

| X | $\Delta_{gem}$ | $\Delta_{cis}$ | $\Delta_{trans}$ | X | $\Delta_{gem}$ | $\Delta_{cis}$ | $\Delta_{trans}$ |
|---|---|---|---|---|---|---|---|
| –H | 0.00 | 0.00 | 0.00 | –CO-OH (isol.) | 0.97 | 1.41 | 0.71 |
| –Alkyl | 0.45 | –0.22 | –0.28 | –CO-OH (conj.) | 0.80 | 0.98 | 0.71 |
| –Alkyl (ring) | 0.69 | –0.25 | –0.28 | –CO-OR (isol.) | 0.80 | 1.18 | 0.55 |
| –C=C (isol.) | 1.00 | –0.09 | –0.23 | –CO-OR (conj.) | 0.78 | 1.01 | 0.46 |
| –C=C (conj.) | 1.24 | 0.02 | –0.05 | –CO-NR$_2$ | 1.37 | 0.98 | 0.46 |
| –Ar | 1.38 | 0.36 | –0.07 | –SR | 1.11 | –0.29 | –0.13 |
| –CH$_2$SR | 0.71 | –0.13 | –0.22 | –SO$_2$ | 1.55 | 1.16 | 0.93 |
| –CH$_2$NR$_2$ | 0.58 | –0.10 | –0.08 | –NR$_2$ | 0.80 | –1.26 | –1.21 |
| –CH$_2$OR | 0.64 | –0.01 | –0.02 | –NCO-R | 2.08 | –0.57 | –0.72 |
| –CH$_2$-Hal | 0.70 | 0.11 | –0.04 | –OR | 1.22 | –1.07 | –1.21 |
| –CH$_2$CO, CH$_2$CN | 0.69 | –0.08 | –0.06 | –OAr, -OC=C | 1.21 | –0.60 | –1.00 |
| –CH$_2$Ar | 1.05 | –0.29 | –0.32 | –OCO-R | 2.11 | –0.35 | –0.64 |
| –CHO | 1.02 | 0.95 | 1.17 | –I | 1.14 | 0.81 | 0.88 |
| –CO-R (isol.) | 1.10 | 1.12 | 0.87 | –Br | 1.07 | 0.45 | 0.55 |
| –CO-R (conj.) | 1.06 | 0.91 | 0.74 | –Cl | 1.08 | 0.18 | 0.13 |

The increments for alkyl (ring) are used when the alkyl group and the alkene are part of the same 5- or 6-membered ring, and those followed by "conj." are used when the substituent or the alkene are further conjugated.

## Tables for estimating $^1$H chemical shifts of mono-substituted benzenes

$\delta\,CH = 7.27 + \Delta$

| X | $\Delta_{ortho}$ | $\Delta_{meta}$ | $\Delta_{para}$ | X | $\Delta_{ortho}$ | $\Delta_{meta}$ | $\Delta_{para}$ |
|---|---|---|---|---|---|---|---|
| –Alkyl | –0.14 | –0.06 | –0.17 | –CHO | 0.56 | 0.22 | 0.29 |
| –C=C | 0.06 | –0.03 | –0.10 | –CO-R | 0.63 | 0.13 | 0.20 |
| –C≡C | 0.15 | –0.02 | –0.01 | –CO-OH | 0.85 | 0.18 | 0.27 |
| –Ar | 0.37 | 0.20 | 0.10 | –CO-OR | 0.71 | 0.10 | 0.20 |
| –SH | –0.08 | –0.16 | –0.22 | –CO-NH$_2$ | 0.61 | 0.10 | 0.17 |
| –SR | –0.08 | –0.10 | –0.24 | –CN | 0.36 | 0.18 | 0.28 |
| –NH$_2$ | –0.75 | –0.25 | –0.65 | –I | 0.39 | –0.21 | 0.00 |
| –NR$_2$ | –0.66 | –0.18 | –0.67 | –Br | 0.18 | –0.08 | –0.04 |
| –OH | –0.56 | –0.12 | –0.45 | –Cl | 0.03 | –0.02 | –0.09 |
| –OR | –0.48 | –0.09 | –0.44 | –F | –0.26 | 0.00 | –0.20 |
| –OAr | –0.29 | –0.05 | –0.23 | –NO$_2$ | 0.95 | 0.26 | 0.38 |

## Chemical shift ranges of $^{13}$C resonances

| Group | $\delta$ range | Group | $\delta$ range | Group | $\delta$ range |
|---|---|---|---|---|---|
| –CH$_2$-, -CH$_3$ | 0–50 | C=C–**CH**R$_2$ | 35–65 | –C=C– | 100–140 |
| C=C–**CH**$_3$ | 5–30 | Hal–**CH**R$_2$ | 40–90 | –C=N– | 140–165 |
| S–**CH**$_3$ | 5–20 | S–**CH**R$_2$ | 55–65 | –C≡C– | 80–105 |
| N–**CH**$_3$ | 15–45 | N–**CH**R$_2$ | 50–70 | –C≡N | 110–125 |
| O–**CH**$_3$ | 50–60 | O–**CH**R$_2$ | 65–90 | Aromatics | 115–150 |
| C=C–**CH**$_2$R | 25–55 | C=C–**C**R$_3$ | 30–50 | Amide C=O | 155–180 |
| Hal-**CH**$_2$R | 20–55 | Hal–**C**R$_3$ | 35–80 | Ester C=O | 155–180 |
| S–**CH**$_2$R | 25–45 | S-**C**R$_3$ | 55–75 | Acid C=O | 170–185 |
| N–**CH**$_2$R | 40–60 | N–**C**R$_3$ | 65–75 | Thioketone C=S | 190–205 |
| O-**CH**$_2$R | 40–80 | O-**C**R$_3$ | 75–85 | Aldehyde C=O | 185–210 |
| | | | | Ketone C=O | 190–220 |

"Hal" is taken to be Br, Cl or F. Iodine can cause alkyl resonances to fall below 0 ppm.

## Appendix 4 Common fragmentations and fragment ions in mass spectrometry

For wider range of data see tables in F. W. McLafferty and F. Turecek, *Interpretation of Mass Spectra*, 4th edn., University Science Books, Mill Valley, CA.

### Common fragmentations

| m/z | Fragment lost | Comment |
|---|---|---|
| M−1, M−2 | H, $H_2$ | |
| M−3→M−14 | | Highest mass peak observed is itself a fragment and not the molecular ion. |
| M−15 | $CH_3$ | |
| M−17 | OH | alcohol, carboxylic acid |
| | $NH_3$ | primary amine (odd MW) |
| M−18 | $H_2O$ | alcohol, aldehyde, ketone |
| M−26 | $C_2H_2$ | |
| | CN | nitrile (odd MW) |
| M−28 | CO, $C_2H_4$, $N_2$ | |
| M−29 | $C_2H_5$, CHO | |
| M−30 | $C_2H_6$ | |
| | NO | nitro or nitroso (odd MW) |
| M−31 | $CH_3O$ | methyl ester or ether |
| M−35/37 | Cl | molecular ions shows two peaks of intensity 3 : 1 two mass units apart |
| M−42 | $CH_2CO$ | methyl ketone or aromatic acetate |
| M−43 | $CH_3CO$ | methyl ketone, acetate |
| M−44 | $CO_2$ | lactone |
| M−45 | $C_2H_5O$ | ethyl ester or ether |
| | $CO_2H$ | carboxylic acid |
| M−46 | $NO_2$ | odd MW |
| M−55 | $C_4H_7$ | butyl ester |
| M−58 | $CH_2=C(OH)CH_3$ | McLafferty rearrangement, methyl ketone with $\gamma$-hydrogen (p. 81) |
| M−60 | $CH_3CO_2H$ | acetate |
| | $HCO_2CH_3$ | tertiary methyl ester |
| | $CH_2NO_2$ | primary nitro (odd MW) |
| M−77 | $C_6H_5$ | monosubstituted aromatic |
| M−79/81 | Br | molecular ion shows two peaks of equal intensity two mass units apart |
| M−81 | $C_5H_5O$ | furylic |
| M−91 | $C_7H_7$ | benzylic |
| | $C_6H_5N$ | *N*-substituted pryridinium (odd MW) |
| M−105 | $C_6H_5CO$ | aromatic ketone or ester |
| M−127 | I | |

### Fragment ions

| m/z | Fragment | Comment |
|---|---|---|
| 15 | $CH_3^+$ | |
| 18 | $H_2O^+$ | |
| 26 | $C_2H_2^+$ | |
| 28 | $CO^+$, $C_2H_4^+$, $CH_2N^+$ | |
| | $N_2^+$· | |
| 29 | $CH_3O^+$, $C_2H_5^+$ | |
| 30 | $CH_2NH_2^+$ | primary amine (odd MW) |
| 31 | $CH_2OH^+$ | primary alcohol |
| 36/38 | $HCl^+$· | |
| 40 | $Ar^+$· | contaminant, useful reference peak |
| 43 | $CH_3CO^+$, $C_3H_7^+$ | |
| 44 | $OCNH_2^+$ | primary amide (odd MW) |
| | $CO_2$ | ester, carboxylic acid |
| | $CH_2CHOH^+$ | aldehyde |
| 45 | $CH_2OCH_3^+$ | ether |
| | $CH_3CHOH^+$ | secondary alcohol |
| 49/51 | $CH_2Cl^+$ | |
| 58 | $CH_2=C(OH)CH_3^+$ | methyl ketone |
| 59 | $CO_2CH_3^+$ | methyl ester |
| | $CH_2C(OH)NH_2^+$ | primary amide (odd MW) |
| | $CH_2OC_2H_5^+$ | ethyl ether |
| 65 | $C_5H_5^+$ | secondary fragment from tropylium (p. 81) |
| 73 | $(CH_3)_3Si^+$ | |
| 77 | $C_6H_5^+$ | monosubstituted aromatic |
| 79/81 | $Br^+$ | |
| 80/82 | $HBr^+$ | |
| 81 | $C_5H_5O^+$ | |
| 85 | $C_5H_9O^+$ | tetrahydropyranyl ether |
| 91 | $C_7H_7$ | tropylium ion, usually the base peak (p. 81) |
| 93/95 | $CH_2Br^+$ | |
| 127 | $I^+$ | |
| 128 | $HI^+$ | |

# Index